A NEW LEAF

A HANDBOOK FOR PRESERVING MICHIGAN'S ENVIRONMENT

A New Leaf is printed
entirely on recycled paper.

Cover and book design
by Mitchell Smith

©1991 Michigan United Conservation Clubs
All rights reserved

ISBN 0-933112-16-5

A NEW LEAF

A HANDBOOK FOR PRESERVING MICHIGAN'S ENVIRONMENT
By Melissa Ramsdell

Edited by Lisa J Allen
Illustrations by Jean MacKenzie

Published by

Michigan United Conservation Clubs
P.O. Box 30235, Lansing, Mich. 48909

CONTENTS

Introduction   i

Chapter 1   1
What are we doing to our water?
From sewage spills to a new, improved Lake Erie, the condition of our waters changes day by day as new threats arise.

Chapter 2   9
Can the lakes survive a spill?
What we transport across our lakes should make us lie awake at night

Chapter 3   15
Protecting the water we have
There's more to water than meets the eye and, therefore, more to worry about. Solutions

Chapter 4   27
Our environment is getting trashed
Recycling, composting and buying less can ease our garbage crunch

Chapter 5   41
The hazards of incineration
A quick fix for trash problems may send clean air up in smoke Solutions

Chapter 6   51
Radioactive waste, toxins and heavy metals
Our past evils are still around to haunt us

Chapter 7   57
Agricultural chemicals may be out of hand
Farmers can turn to methods that are kinder to the land

Chapter 8 — 65
Dangers at home
Homes may be toxic dumps if we're not careful about what we buy and throw out

Chapter 9 — 73
Invisible Dangers
Air serves as a highway for dangerous pollutants

Chapter 10 — 81
Indoor Air
Good ventilation is an easy cure for chronic problem

Chapter 11 — 87
How does Michigan penalize polluters?
Our state isn't always as tough as it ought to be
Solutions

Chapter 12 — 97
Habitat Destruction
Wildlife pays the price for our quest for new horizons
Solutions

Chapter 13 — 111
Are we destroying our chance for survival?
Every act carries global implications

Chapter 14 — 123
Forests can reverse the damage
Trees may counter humans' reckless behavior
Solutions

Conclusion — 131
What is a consumer to do?
Turning "green" starts at the store

Epilogue — 135

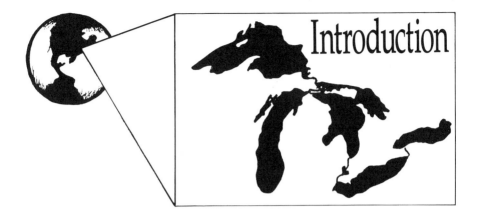

Introduction

Michigan is a place of great natural beauty. Clearly, people who live here value their pristine surroundings. Michigan has one of the nation's largest holdings of state forests and parks and can boast about having more campsites than any other state. Situated as the focal point of the Great Lakes basin, Michigan is bordered by 3,288 miles of sandy coastline and is dotted with more than 11,000 inland lakes. As a result, we have more boat and vacation property owners than practically any other state. Michigan's travel, hunting, fishing and tourism trade is a multibillion dollar industry.

Michigan and the Great Lakes basin comprise a unique ecosystem. As the largest interconnected system of freshwater lakes in the world, the Great Lakes represent one-fifth of all of the fresh surface water on earth and 90 percent of the available fresh surface water in the United States. More than 20 million people rely upon the Great Lakes for their drinking water.

Because so many lives depend on our natural resources, they must be responsibly managed. We must be stalwart stewards of the natural wealth that our home provides. After all, our lives depend on it.

Michigan United Conservation Clubs (MUCC), with 140,000 members, is the largest conservation group in Michigan. This book represents our belief in the power of individuals to save our state and the earth from environmental degradation. It is hoped this guidebook will help you recognize that your everyday choices count toward making Michigan a better place to live.

We've divided the book into five parts, each addressing an aspect of our surroundings. At the end of each part, we've listed a variety of things you can do at home and in your community to protect our vital resources. We have provided lists of key phone numbers, conservation organizations, government officials and helpful things to read.

Of course, we understand that people have different levels of commitment to the environment. Starting simply, changing the way you do some things at home may deepen your commitment and inspire you to join an ecological organization, take an active role in your community (by starting a local recycling center, for example) or try to influence environmental legislation. Throughout the 1990s, the enormous ecological challenges we must grapple with will require a strong sense of personal responsibility and greater involvement.

Michigan citizens have already shown that they are willing to take action to preserve their unique environment. In 1976, more people voted to impose a bottle deposit than voted for presidential candidates. The referendum, started and led by MUCC, placed a mandatory deposit on beer and soft-drink containers to reduce roadside litter and encourage recycling. Once it took effect in 1978, people had to change their life-styles a bit, but they now are pleased with the benefits those changes brought for Michigan's environment.

Ten years after the bottle bill took effect, Michigan voters turned out once again in favor of the environment. They authorized the sale of state bonds to spend $660 million to clean up toxic waste sites, upgrade city sewer systems and to fund recycling and Great Lakes environmental research projects. Another $140 million in bonds will enhance state and local parks.

At the local level, people are banding together to keep their homes safe. Residents of Lenawee County taxed themselves to pay an attorney to oppose a proposal to place a low-level radioactive waste dump there. Voters routinely support programs to increase recycling, composting and household hazardous waste collections.

The next step is to move environmental awareness from the voting booth to our everyday lives. Ideally, everyone soon will have trash bins with separate compartments for recyclables and cloth bags for their groceries. Shoppers will use their economic clout to encourage supermarkets and eating places to reduce packaging and provide more recycled and environmentally-sound products. Maybe someday automakers will sell solar-powered electrical cars that don't emit carbon and nitrogen oxide. A Warren resident already has developed a solar-powered lawn mower.

It is hoped this book will leave you feeling both disturbed by the mess we've made of our state's precious resources in just a few generations and inspired by the challenge we've been given to restore Michigan to its former glory. That way many generations can enjoy the natural wonders we've come to expect in Michigan.

Introduction / **iii**

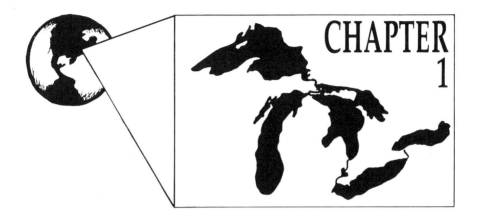

What Are We Doing To Our Water?

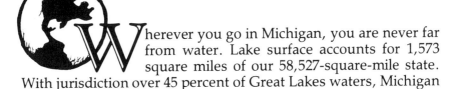herever you go in Michigan, you are never far from water. Lake surface accounts for 1,573 square miles of our 58,527-square-mile state. With jurisdiction over 45 percent of Great Lakes waters, Michigan has a special role as a leader in protecting this vast resource which dictates much of our quality of life.

The lakes represent a significant asset to the state's economy because they provide water for industrial manufacturing, energy production, shipping, fishing, recreation and tourism. Also, more than half of all Michigan residents rely on the lakes for drinking water.

But many of our lakes and rivers don't get the respect they deserve. Many waterways and lakes are used as open gutters for municipal and industrial wastes. Every year, 16 billion to 20 billion gallons of untreated sewage and industrial waste are dumped into Michigan waters because sewage treatment facilities are overwhelmed when it rains.

In Michigan, many communities use the same pipe system for sewage and rainwater runoff. When it rains, too much fluid flows to treatment plants. Some of the excess is diverted to pipes that dump the effluent directly into public waters. This occurrence is called combined sewage overflows and they occur in nearly 100 Michigan communities. It will cost $2 billion to $3 billion to adopt systems that will marginally treat all fluid during rainstorms.

The Detroit Wastewater Treatment Plant, the largest discharger of untreated waste into the Great Lakes, treats approximately 920 million gallons of wastewater per day. The facility, which accommodates nearly 70 communities, is sometimes overloaded when as little as a quarter-inch of rain falls in some areas. The state Water Resources Commission ordered the City of Detroit to improve its facilities to handle a 1.7-inch rainfall in a one-hour period. The commission said the city should be able to retain the water, allowing solids to settle and enabling substances to be skimmed from the top. All discharges also must be disinfected before release, the commission said.

The city has contested the order, arguing that the requirements are too strict. MUCC and the National Wildlife Federation entered the case, saying the requirements are too lax. The dispute will drag on for years. Meanwhile, billions of gallons of untreated sewage from Detroit and other communities around the state will continue to flow into the lakes.

But all is not bleak. Some things have been improving in the Great

Lakes basin since the 1960s, when the first environmental regulations for controlling direct discharges of water pollution were put into place. Cutting down on the amount of phosphate detergents being poured into the lakes was a leading priority. Phosphate, a fertilizer, is used in detergents to soften water. It stimulates explosive growths of algae in the water, which consumes most of the oxygen and suffocates fish. Laundry and soap detergent companies have been forced to decrease their use of the chemical.

Lake Erie, once called the "American Dead Sea" because it was so polluted, has been restored to a cleaner state. Today, a walleye fishery flourishes there.

Ohio's flaming river, the Cuyahoga, no longer catches on fire as it did in the 1960s. The river ignited because of the amount of flammable chemicals dumped into it by the industries along its Cleveland banks. Although it still contains contaminants, the river is now considerably cleaner. These two success stories resulted from a significant effort to decrease pollution discharges during the 1970s.

Despite this progress, persistent toxic pollutants like man-made organic or carbon-based agricultural chemicals, mercury and lead still taint the Great Lakes. It is frightening to note that approximately 1,000 toxic chemicals—362 of them in large amounts—have been found in Great Lakes water, fish and wildlife. Some of these substances, which do not readily break down in water, can persist for decades and are extremely poisonous to humans and wildlife, even in trace amounts. Although these threatening chemicals are invisible, they have the potential to increase the risk of cancer, genetic mutations and birth defects through continual exposure. Recent studies have shown that some fish and bird species living in the Great Lakes basin are already manifesting these symptoms.

The federal Environmental Protection Agency has warned people not to eat large lake trout from Lake Superior because they contain potentially dangerous amounts of PCBs in their skin and fat. (PCBs, or polychlorinated biphenyls, are used primarily in electrical transformers.) The Michigan Department of Public Health routinely issues fish advisories for most inland waters and all of the Great Lakes.

Cancerous tumors have been found in some species of Great Lakes fish and beluga whales swimming near the mouth of the St. Lawrence River. Chemicals hinder reproduction of lake trout as well as the herring gulls and bald eagles that nest near the shoreline. Genetic mutations, such as club feet, crossed beaks and missing eyes, have been discovered in cormorants that feed on contaminated fish.

Where are all of these poisonous pollutants coming from? It is often difficult to track them to their origin. Some hazardous substances enter the lakes from municipal combined sewage overflows. Others are still there from the days when industries poured their waste directly into lakes and rivers. Federal law now requires that corporations treat their wastes, but dumping continues where regulations are not strictly enforced.

Direct discharges, or point sources of contamination, aren't all we have to worry about. Almost 25 percent of all water contaminants are carried to water via air, rain or snow. About 80 percent of the pollution in Lake Superior and half of the pollution in lakes Michigan and Huron comes from the air. Our greatest challenge will be to minimize the toxins entering the water from indirect, or non-point sources. It is a subject that will be discussed in a later chapter.

Other non-point sources of water pollution include leaking landfills, chemically-treated farm fields and malfunctioning septic tanks. Toxic chemicals and harmful bacteria can seep into the groundwater and travel to lakes, streams and rivers. Agricultural fertilizer can trigger explosive weed and algae growth that jeopardizes other aquatic life.

Hazardous waste disposal sites are another source of indirect contamination of the groundwater. Before the 1979 Hazardous Waste Management Act, hundreds of small, rudimentary hazardous waste treatment sites were spread across the state. Often, companies would simply dig a hole in the ground and dump their wastes into an unlined pit. Hazardous substances oozed from the pit into the groundwater and eventually found their way into the lakes. While modern hazardous waste treatment facilities are much more sophisticated, none of the current technology guarantees long-term protection from leaks.

Even if we could eventually eliminate all new sources of water pollution, existing contaminants will continue to mix in water, settle into the sediment on the bottom and cycle through the biological food chain. Contaminated sediments stirred up from the bottom of lakes by dredging, ships, fish or storms reintroduce toxic contaminants into the water. Substances that do not easily break down in water, like PCB and DDT, concentrate in the organisms that ingest them.

For instance, when persistent organic chemicals such as PCBs enter the water, they become suspended in water molecules and are consumed by tiny plant and animal organisms called phytoplankton and zooplankton. The smallest life forms in the ecosystem, they represent the bottom of the aquatic food chain. When small fish such as alewives and young trout consume thousands of zooplankton, the toxic substances present in the zooplankton accumulate in the fish's fatty tissue. The process continues as the toxic chemicals ascend the food chain into large predator fish such as lake trout and salmon, as well as several species of fish-eating birds like herring gulls.

Due to this process, called biological magnification, toxic substances that are barely detectable in water may be present in large fish in concentrations that are a million times greater. Humans sit at the top of the food chain, so if we eat Great Lakes fish we are at a much greater risk than if we drink the water or use it for bathing, swimming or recreation. Researchers still are unsure what frequency of Great Lakes fish consumption is safe and how much might pose a threat to human health. A 1984 study by Wayne State University surveyed more than 200 women who ate an average of two or three meals of Lake Michigan fish per month while pregnant.

The women in the study had higher concentrations of PCBs in their body tissue and breast milk than mothers who did not regularly eat Lake Michigan fish. They had shorter pregnancies and their babies were born with lower than average birth weights, smaller head circumferences and some neuromuscular abnormalities. A follow-up study examined the children four years later and found that the toxins may be affecting their short-term memories and hindering their ability to develop math and reading skills. The study pointed out that because PCBs are passed from mother to child, measurable quantities of PCB will be passed onto the next five generations.

Clearly, there is a great need for more long-term research projects and more consistent government advisories about the health effects of eating the fish in our region. While many questions remain, there is no doubt that the health risks will increase if we continue to add more toxic chemicals to the water. If we do not stop the pollution coming into our lakes and clean up what's already there, Michigan's thriving and profitable sport-fishing industry will be threatened.

Because the Great Lakes seem almost as vast as an ocean, people once thought they could dump almost any amount of waste into them without causing any environmental harm. But unlike an ocean, the lakes are a closed system that flushes out water very slowly. It takes 2.6 years for the water in Lake Erie to be replaced; 6 years for Lake Ontario; 22.6 years for Lake Huron; 99.1 years for Lake Michigan; and 191 years for Lake Superior. Simply put, if someone dumps some hazardous waste into Lake Superior today, it will remain there until long after we are dead.

Contaminants in the Great Lakes are particularly dangerous because the lakes are a closed system—a huge bathtub—lacking the natural cleansing activity of the ocean, which helps disperse toxic substances. And unlike salty ocean waterways, there are more than

60 drinking water intakes, serving millions of people, along Michigan's freshwater shoreline which must be protected from contamination.

With 1,000 toxic substances already present in the lakes and more flowing in from contaminated tributaries every day, all new sources of contamination — from the air, land or water — need to be stopped so officials can concentrate on cleaning up those that already exist. This goal of "zero discharge" of hazardous substances in toxic concentrations is the philosophy of the Great Lakes Water Quality Agreement between Canada and the United States. It was signed in 1972, redrafted in 1978 to include toxins and amended in 1987.

The agreement calls for the "virtual elimination" of persistent toxic substances in the Great Lakes basin. The International Joint Commission, formed under the 1909 Boundary Waters Treaty as a governing body to oversee the lakes shared by the United States and Canada, recently recommended using Lake Superior as a pilot project for zero discharge. However, American and Canadian governments are slow to implement it. The IJC described the goal of virtual elimination as an "unmet challenge," but some Great Lakes citizens groups are calling it an unfulfilled promise. In fact, a 1989 General Accounting Office study found that few recommendations of the international panel have ever been acted upon.

But as hazardous chemicals pose an increasing danger to the fish and other wildlife that live in the lakes, as well as the people who consume them, eliminating these substances is essential.

But state and provincial governments and agencies such as the IJC and the EPA do not bear the entire responsibility of making sure that zero discharge becomes a reality. As residents of the Great Lakes basin, we have a duty to learn about the environmental consequences of our daily habits and to become more involved in government as advocates of cleaner Great Lakes water. The goals in the international water quality agreement cannot be achieved unless citizens push state and federal governments to meet them.

Another problem threatening Michigan's water supply is the introduction of exotic species carried into the Great Lakes by ships that have been traveling in foreign waters. A recent and dramatic example is the arrival of the now ubiquitous mollusk species, the zebra mussel. Named for its striped shell, the mussel has been attaching itself to most hard surfaces in the water and reproducing rapidly. A healthy female produces about 40,000 eggs a year. The prolific mollusks wreaked havoc in the City of Monroe by plugging up its municipal water intake pipe. They also foul up boat engines by getting into intake pipes and have caused problems for some power plants along Lake Erie.

The mussels were supposedly picked up by ships in Europe when they took in ballast water in place of cargo. Once they arrived in the Great Lakes, the ships let out their ballast water to make room for cargo, unleashing hordes of mollusks. The mussels have made their way into all of the Great Lakes and some inland waters.

Zebra mussels can be removed from intake pipes with a high-pressure spray of water, chlorine, hydrogen peroxide or molluskicides. However, some environmentalists are concerned that if municipalities begin flushing large amounts of chlorine through their water intake pipes to exterminate the mussels, they will add yet another source of pollution into the Great Lakes. This problem has no easy cure.

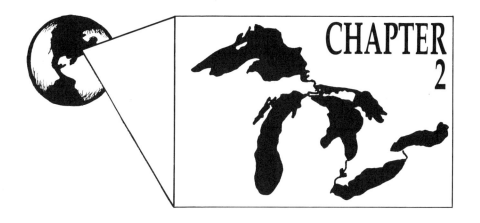

Can The Lakes Survive A Spill?

The disastrous Exxon oil spill in Alaska's Prince William Sound and the wartime unleashing of millions of gallons of oil into the Persian Gulf have raised some fears about the large ships that pass through the Great Lakes.

The governors of the eight Great Lakes states agreed in 1986 to ban oil drilling in the U.S. portion of the basin, lessening the threat of oil contamination. The product, crude oil, is not transported on the Great Lakes but is shipped through transcontinental pipelines.

But other threats persist. Refined petroleum products such as gasoline, kerosene, jet fuel and heating oil routinely float across the lakes. Shipments of hazardous chemicals including benzene, materials to make plastics, fertilizers, pesticides, paints and solvents are regularly carried across Great Lakes waters.

While the transportation of these materials is definitely risky, the ships that travel the Great Lakes are nowhere near the size of supertankers like the Exxon Valdez. The Valdez, which can hold 52 million gallons, spilled more than 10 million gallons of crude oil. In comparison, the largest ship on the Great Lakes is a Canadian vessel which carries about five million gallons. The largest U.S. vessel holds three million gallons.

Ninety-five percent of the spills on the Great Lakes are less than 2,500 gallons, whereas the EPA defines a major spill as more than 10,000 gallons and a catastrophic spill as 200,000 gallons or more of oil or hazardous chemicals.

However, the regulation of hazardous substance shipping varies between the United States and Canada. The five U.S. vessels that carry hazardous substances on the Great Lakes must have protective double hulls, but there is no such requirement for the 33 Canadian-owned vessels. Plus, the U.S. Coast Guard has a tracking system that regulates vessel traffic and responds to any problems ships may have. American Great Lakes vessel pilots are required to go through an extensive training program and the Coast Guard requires that all foreign vessels have on board a trained Great Lakes captain who can take over if necessary as the ships ply the often treacherous waters of the Great Lakes. These rules are less stringent in Canadian waters.

The potential for disaster is ever present. Shortly after the Exxon Valdez spill, Michigan nearly experienced a catastrophic spill of its own. On March 30, 1989, a Canadian tanker carrying 1.4 million gallons of carbon black feedstock, a toxic, carcinogenic substance, almost ran aground twice in the Detroit River. The vessel came perilously close to spilling its contents near the north end of Belle Isle, the site of the drinking water intake that supplies water for approximately three million people—about one-third of Michigan's entire population.

A spill was avoided when the U.S. Coast Guard ordered the ship to release some of its ballast water to raise it to a safer level in the water. A spill would have been disastrous—most of the state's spill response experts were in Alaska cleaning up the Valdez mess.

In another incident, 176,000 pounds of styrene, a hazardous chemical used to make plastic foam and other materials, was accidentally released from a Dow Chemical Company plant in Sarnia, Ontario. Into the St. Clair River flowed an amount of styrene that equaled the facility's legal release limit for the next 1,400 years!

Incidents like these are not uncommon in Michigan. The U.S. Coast Guard's Detroit office alone reported an average of one oil or chemical spill every four days in 1988. Spills of acid, paint, styrene and benzene are a common occurrence along the Canadian side of the St. Clair River. The Great Lakes absorbed more than 5,000 oil and hazardous chemical spills during the 1980s and only 20 percent came from vessels, according to the U.S. Coast Guard.

When spills do occur, the Coast Guard—which is given the authority to deal with spills by the Clean Water Act—relies heavily on private contractors to provide cleanup equipment, but much of it is in poor condition and unable to withstand battering Great Lakes waves.

The Michigan Department of Natural Resources has a fleet of about 50 privately-owned boats available for spill response around the state but no centers exist near the Great Lakes to rescue wildlife from oil or hazardous chemical spills. Containment equipment is not readily available either to protect these vulnerable, environmentally sensitive areas.

There are only a handful of people in Michigan who are trained to handle hazardous materials in an emergency, but that is changing. A hazardous materials training center is underway at Michigan State Police headquarters in Lansing as part of its Emergency Management Division. The $350,000 center, donated by the Michigan Chemical Council, transporters and petroleum companies, will begin training in June 1991. It will provide inexpensive training for local emergency professionals handling oil and hazardous spills on waterways.

Funding for more spill containment equipment is badly needed. Once these key components are in place, trained personnel can participate in frequent, unannounced simulated spills to practice communication and cleanup procedures. This way, they can work most of the bugs out of the system before disaster strikes.

Michigan is not alone in its lack of preparation for a spill. Experts say none of the states is prepared to contain a major spill. The entire basin is woefully short of funding for cleanup equipment, wildlife protection, trained workers and coordinated emergency response plans. But maybe the Valdez was a blessing in disguise by jolting us into the realization that we must beef up our prevention and cleanup capabilities if we are to protect the Great Lakes basin and its inhabitants from a similar accident. With more shoreline than any other state except Alaska and the country's largest supply of fresh water, Michigan has the most to gain from a particularly vigilant spill prevention and preparedness plan.

But how are we going to prevent the oil spills that happen every day? People who change their own automobile oil and then dump it down the sewer, spread it over a dirt road or take it to the curb in their garbage can are collectively creating an oil slick that is larger than the 10.9 million-gallon Exxon Valdez oil spill, according to a study by the West Michigan Environmental Action Council. In fact, absorbent oil booms like those that were used to clean up Prince William Sound are usually placed along the edge of the Rouge River to collect the annual oil runoff during spring thaw and heavy rains.

Oil floating over water impacts aquatic life by blocking out sunlight, killing the tiny organisms that support the rest of the aquatic food chain. The lead and other harmful substances that accumulate in automobile oil during the engine's combustion process dissolve in water and contaminate fish and plants and poison drinking water supplies. If those chemicals are ingested by humans and animals, they can cause serious health problems, including cancer.

An estimated 26 million gallons of motor oil are used each year in Michigan and half of it is purchased by do-it-yourself oil changers. Currently, 11 percent of this oil is taken to 700 different collection sites at community recycling centers, government garages, local gas stations, instant oil change facilities or chain stores that sell auto parts. Most garages store used oil in large underground tanks or above-ground barrels. A used-oil dealer will purchase the oil from the garage and haul it away to sell to an oil re-refiner.

Unlike gasoline which is burned up to fuel the engine, engine lubricating oil does not get used up in the combustion process, it just gets dirty. In order to be reused, the lead and other pollutants must be removed. But it is worth it. Recycling used motor oil is much cheaper than making it from fresh crude oil and it saves our limited reserves of fossil fuels. A 42-gallon barrel of crude oil will yield only about 2 ½ quarts of virgin motor oil. It takes only one gallon of used motor oil to make the same amount by recycling.

Processed motor oil also can be burned as fuel for electricity or other uses.

Although it appears to be a small fraction, Michigan's 11 percent used oil recovery rate is one of the highest in the nation. The convenience of oil collection centers and increased education about the environmental and health impacts of improper oil disposal can bring the state closer to eliminating oil contamination. Retailers and consumers also can play an important role by creating a market for the product by supplying and buying re-refined motor oil.

The Michigan Legislature recently approved legislation to make dumping used oil down a drain or into the trash a misdemeanor. It also requires the Department of Natural Resources to establish and coordinate recycled oil collection sites throughout the state and develop a funding plan.

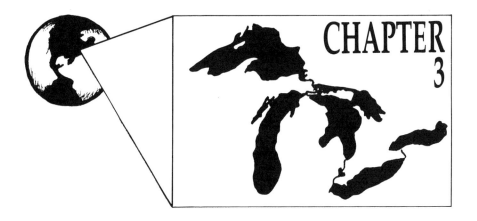

Protecting The Water We Have

The reason for all of our water protection programs is obvious. We're standing on it. Underneath the soil lies our "sixth Great Lake"—groundwater. The total volume of Michigan's groundwater supply is said to be four times the volume of all of the Great Lakes combined.

Although it has been called a sixth lake because of its size, Michigan's underground aquifer actually works more like a sponge, soaking up and storing the water that falls to the ground via rain or snow. Usually, the water is stored in the tiny spaces and cracks between rocks and soil deep beneath the earth's surface. In a few places, there are networks of underground lakes and streams and when water levels begin to decrease above ground, groundwater replenishes surface streams, wetlands, lakes and rivers. In heavy rains, groundwater prevents floods by absorbing some of the water.

Nearly half of Michigan's 9 million residents—mostly in the southern part of the state—depend on this hidden water resource for drinking water. Thirty-seven percent of farmers use groundwater for irrigation and livestock water. Others who depend upon Michigan groundwater include industries, home gardeners, golf courses and city parks.

Water under the ground is especially vulnerable to contamination because it moves very slowly—only a few inches to a few feet per day. Once a pollutant arrives, it is likely to remain in groundwater for a long time. Also, groundwater has less oxygen and heat than surface water does. These are key ingredients needed to break down harmful substances. Once contaminated, it is very difficult and costly to clean it up.

There are many sources of groundwater contamination in Michigan. Some of the most common culprits include improperly lined municipal and industrial landfills, ponds or lagoons; malfunctioning septic tanks and city sewer lines; road salt; and agricultural fertilizers and pesticides. Farm manure, which makes its way into drinking water supplies via groundwater, carries bacteria that can spread hepatitis and other diseases if not properly treated.

Other sources of groundwater pollution include leaking underground gasoline, oil or chemical tanks that have been poorly maintained or abandoned altogether. The spills caused by truck and train accidents sometimes carry hazardous materials to groundwater, too. Home owners add to the problem by pouring motor oil, gasoline and antifreeze onto the ground and using excessive amounts of fertilizers and insecticides.

The DNR has identified more than 2,000 contaminated groundwater sites in Michigan and about 240 new sites are being added to the list every year. Because groundwater contamination is so difficult to clean up, preventing the pollution before it reaches the ground is crucial.

While one of Michigan's major strengths is the seemingly unlimited freshwater supply, perhaps our greatest weakness has been our inability to keep that water supply clean and pollution-free for drinking, recreational enjoyment and habitat preservation. Even though we are surrounded by ample bodies of water, we must not allow ourselves to be lulled into a false sense of security that our water supply will never run out, even if we are careless and wasteful. While the Great Lakes are not likely to dry up in our lifetime, our activities could render their waters unfit for consumption.

One strong step in the right direction is to use less water and thereby reduce its contamination. Water conservation is vital if we want to preserve the quality of the water. Installing low-water toilet fixtures or shower heads and reducing general water use in the home can help conserve limited groundwater supplies. This becomes especially important as more and more underground drinking water becomes contaminated and unfit for human consumption.

Another threat to our water resources is theft. Citizens from areas that have nearly exhausted their water supply may cast an envious eye upon our unique freshwater basin. This thirst for diversion will most likely occur when global warming and its accompanying higher temperatures cause water levels to drop. Meanwhile, anticipated population growth in the Great Lakes basin could cause water consumption to double between 1975 and 2000 and triple from 2000 to 2035.

The Ogallala Aquifer, an 800-mile-long underground body of sand and water which supplies groundwater for the southwest Great Plains states (Texas, Nebraska, Kansas, Oklahoma, Colorado and New Mexico), is expected to run out of water by 2020 at its current rate of use. Careless agricultural irrigation practices that cause huge amounts of water to evaporate or drain off are drawing out water faster than nature can replace it. Approximately 200,000 wells tap into the immense aquifer to irrigate 16 million acres of the country's most productive farmland. Without the water resources to irrigate this area, the nation's food supply could be threatened.

Proponents of diversion in the West have already begun looking into tapping into the Great Lakes. One project proposes transporting the water through a system of canals from Lake Michigan through Chicago and into the West through the Mississippi and Missouri rivers. The water also could be diverted from Lake

Superior in Minnesota to the Missouri River.

If diversion of Great Lakes water to the Southwest by way of giant pipelines succeeds, it would spell environmental and economic disaster for the Great Lakes basin. Such a large scale diversion would lower the Great Lakes water levels and significantly lower coastal wetland areas. Wetlands provide nursery grounds for fish and wildlife and act as a natural water cleanser.

The lowered lake levels would limit the size of the ships that could travel the waters and the amount of cargo that vessels could safely carry, severely affecting the region's economic well-being. The Army Corps of Engineers peg losses at $105 million per year for shipping and hydroelectric power if water is diverted. As one Detroit-area water specialist aptly observed, "When you export water, you export the economy that goes with it."

Diversion also would affect Great Lakes drinking water quality. Twenty-four million Americans and Canadians use the Great Lakes as a source of drinking water. As the region's population grows and more groundwater supplies are contaminated, the number of communities tapping into lake water for drinking supplies will increase. Lower water levels resulting from diversion and increased population pressure will likely cause shoals of contaminated sediment to be exposed to wave action, stirring up from the bottom PCB, DDT, chloradane and toxaphene. These extremely dangerous substances are difficult to remove if they mix with municipal drinking water supplies.

Lower lake levels interfere with the natural flux in water levels and stimulate excessive algae growth, which infringes on fish spawning areas and natural marshes. Because so much of the inherent beauty of Michigan depends on water, our booming tourism, travel and fishing business would dry up along with the lakes. But most importantly, quality of life would decline for the people and wildlife who depend upon this life-sustaining resource.

We need international cooperation to protect the Great Lakes. The Great Lakes basin is shared by 40 million people who represent two nations, eight states, two provinces and countless cities and counties. We must work together to protect our shared asset—a fragile and complex ecosystem—from such threats as diversion, toxic contamination and the loss of wildlife. Wise management of our common resources will require greater public awareness and a stronger political will to protect the lakes. If we succeed, the Great Lakes region will set an example for the rest of the world in international cooperation for the benefit of the environment.

As stewards of our state's most vital resource—water—we need to continue keeping a close watch on the ends of pipes to make sure industries along the lakes and rivers are responsible. Michigan residents need to be better educated about their daily impact on groundwater. Our universities must increase Great Lakes research so we can better understand the health risks that are floating in our waters and the government needs to improve the quality and consistency of fish consumption advisories so the fishing industry knows where it stands.

Our emphasis has to shift to prevention and anticipation instead of crisis management. We must make water quality preservation a top environmental priority in both our domestic and civic lives. If we can make these changes, we can be sure that we leave the Great Lakes as a legacy rather than a burden for future generations.

SOLUTIONS
Things to do:

• Take quick showers rather than baths; they use less water. A water-saving shower head can save 50 gallons of water for every 10-minute shower you take.

• Don't let the water run while you shave or brush your teeth.

• Wait until you have a full load to run the dishwasher and washing machine.

• Use phosphate-free soap and detergent. Phosphates encourage algae growth which degrades water quality.

• Equip your toilet with a space occupier to save water and if you're buying a new one, get a water-saving model. Check your toilet for leaks and repair them. A leaking toilet can waste up to 200 gallons of water per day.

• Install a low-flow aerator in your sink faucets. Change worn out washers to stop dripping faucets, which waste hundreds of gallons of water each week.

• If you have an older home, test your water for lead.

• Learn about Great Lakes fish contamination. Be prudent about the amount of Great Lakes fish you and your family eat.

• Read your water meter once a week and compare past and present water bills to gauge the success of your conservation efforts.

• Wash your car with a bucket of soapy water instead of letting the hose run.

• Water your yard in the morning or evening to minimize evaporation.

• Try kitty litter or sand on your sidewalks in the winter. Salt can damage vegetation and kill aquatic organisms if it makes its way into a pond or stream.

• Plant native, drought-resistant plants in your yard.

• Use mulches or compost to retain water around your garden and minimize watering.

• Cut your grass higher. Two- to three-inch blades make the lawn more drought-resistant and encourage healthy roots. Or replace portions of your lawn with meadow, walkways or decks.

• Recycle your motor oil or make sure the place that does it for you recycles the oil. One quart of oil can pollute 250,000 gallons of drinking water.

• Test your well for contamination by hazardous chemicals and harmful bacteria—especially if you have a septic tank or live near frequently fertilized or sprayed cropland. Have your septic system checked periodically and repair it if necessary.

Things to do in your community:

• Learn about local, state and federal anti-water pollution laws. Find out how they affect you as a resident of the Great Lakes basin and what needs to be changed.

• Write letters, attend meetings and join citizens' groups to pressure state government to get polluters to clean up their messes.

• Join a citizens' group that is actively working on the water pollution issues that interest you.

• Attend International Joint Commission (IJC) public meetings, hearings or workshops on Great Lakes pollution. See that its recommendations are implemented.

• Help restore the IJC's 42 Areas of Concern by working on your community's remedial action plan.

• Make sure fish consumption advisories are consistently available to the public and are up-to-date.

- Encourage school officials to test drinking fountains for lead and ask them to purchase water-saving and energy-conserving toilets, faucets and water heaters.

- Make sure your public library has a collection of books on water pollution and the Great Lakes. If not, help them develop one.

- Urge local government, schools and businesses to substitute sand or kitty litter for salt during the winter.

- Request that your local water supplier check for and repair leaks in delivery pipes.

Contacts:

- Michigan Department of Natural Resources Surface Water Quality Division. Contact for questions about Michigan water pollution. (517) 373-1949.

- Groundwater Education in Michigan, a joint effort between the MSU Institute for Water Research and the W.K. Kellogg Foundation. It sponsors studies, workshops and education programs about Michigan groundwater contamination. 334 Natural Resources Building, Michigan State University, East Lansing, Mich. 48824-1222, (517) 353-3742.

- Michigan Environmental Council can provide information about groundwater contamination. 115 West Allegan, Suite 10B, Lansing, Mich. 48933, (517) 487-9539.

- East Michigan Environmental Action Council (EMAC) trains volunteers to carry out groundwater protection and education projects. 21220 West 14 Mile Road, Birmingham, Mich. 48010, (313) 258-5188.

- For information about recycling used motor oil: West Michigan Environmental Action Council, 1324 Lake Drive SE, Grand Rapids, Mich. 49506, (800) 634-9504; and the Michigan State University Cooperative Extension Service, East Lansing, Mich. 48824-1222, (517) 355-2308.

- Great Lakes Natural Resource Center, National Wildlife Federation, provides free publications such as "Lake Michigan Fish: Should You Eat Your Catch?" The center also provides a series of pamphlets with information about zero discharge of toxic substances into the Great Lakes. 802 Monroe, Ann Arbor, Mich. 48104, (313) 769-3351.

• Lake Michigan Federation is a group devoted to protecting Lake Michigan from pollution and environmental degradation. It publishes a booklet, "Great Lakes Toxic Hotspots: A Citizen's Action Guide," which contains specific steps for developing and implementing a cleanup plan for a toxic hot spot near you. 8 South Michigan Ave., Suite 2010, Chicago, Ill. 60603, (312) 263-5550.

• Great Lakes United, Cassity Hall, State College at Buffalo, is an international coalition of more than 1,000 citizens concerned with protecting the Great Lakes-St. Lawrence basin. It publishes a booklet, "A Citizens' Guide to the Great Lakes Water Quality Agreement," which gives practical explanations of the IJC and the complex web of legislation and international agreements involved in protecting the Great Lakes basin. 1300 Elmwood Ave., Buffalo, N.Y. 14222, (716) 886-0142.

• For information about home water quality, contact the Water Quality Association, Box 606, Lisle, Ill. 60532.

• The National Xeriscape Council can provide information about water-efficient landscape design, 8080 South Holly, Littleton, Colo. 80122.

• To report a hazardous spill in the Great Lakes, call the U.S. Coast Guard National Emergency Response Center, 1-800-424-8802, or the Michigan Department of Natural Resources Pollution Emergency Alerting System, 1-800-292-4706.

• Environmental Protection Agency's Safe Drinking Water Hot Line: 1-800-426-4791.

• International Joint Commission provides free pamphlets about many Great Lakes water quality issues. The agency publishes biennial reports on the state of the Great Lakes and holds many meetings and hearings that are open to the public. Great Lakes Regional Office, 100 Ouellette Ave., 8th Floor, Windsor, Ontario N9A 6T3, (519) 256-7821, or Box 32869, Detroit, Mich. 48232, (313) 226-2170.

Further Reading:

IJC publications:
 • "The IJC: What it is, How it Works" pamphlet describes the IJC's

role in protecting the Great Lakes.

• "Focus," the IJC's quarterly publication, provides news about Great Lakes water quality, IJC activities, upcoming events and resource materials.

• "The Great Lakes, a Vital Resource Worth Protecting" discusses the ecosystem approach to protecting the Great Lakes, the Great Lakes Water Quality Agreement, and the various functions of the IJC.

• "A Citizen's Guide to Great Lakes Pollution Problems and Solutions" offers a comprehensive discussion of the sources of Great Lakes pollution and suggests ways concerned citizens can help protect the lakes.

• "Remedial Action Plans For Areas of Concern" describes the process for developing and implementing a remedial action plan for a polluted section of the Great Lakes basin.

• "Toxic Substances" discusses the toxic chemicals that are manufactured in the Great Lakes basin and have been found contaminating Great Lakes waters. It explains government efforts to clean up and control toxins and how citizens can participate in solving the problem.

• "Great Lakes Diversions and Consumptive Uses." This 80-page report describes existing diversion projects, various large-scale uses of Great Lakes water and proposed new diversion projects.

• "Directory of Great Lakes Education Material" provides a listing of more than 250 materials that are available to teachers.

Six fact sheets on the Great Lakes are available through the Michigan Sea Grant Extension, Cooperative Extension Service, Michigan State University, 10-B Agriculture Hall, East Lansing, Mich. 48824-1039, (517) 355-0240.

Great Lakes, Great Legacy? by the Conservation Foundation of Washington, D.C., and the Institute for Public Policy in Ottawa, Ontario. It covers toxic contamination's health threat to humans and wildlife in the Great Lakes. Write The Conservation Foundation, Box 4866, Hampden Post Office, Baltimore, MD 21211, or call (301) 338-6951.

"The Great Lakes: a Stewardship Left Untended," a special section reprinted from the 1988 annual report of the Charles Stewart Mott Foundation. It concerns environmental threats to the Great Lakes and how citizens are working to protect them. Charles Stewart Mott Foundation, 1200 Mott Foundation Building, Flint, Mich. 48502-1851, (313) 238-5651.

"Citizen's Guide to Drinking Water and Citizen's Guide to Community Water Conservation" is free from the National Wildlife Federation, 1400 16th St. NW, Washington, D.C. 20036.

Is Your Water Safe to Drink? by Consumer Reports Books (1988).

"Lead and Your Drinking Water," free from the Environmental Protection Agency, Office of Public Affairs, Washington, D.C. 20046.

"Lead in School Drinking Water" is available through the Government Printing Office, Dept. 36ES, Superintendent of Documents, Washington, D.C., 20402-9325, (202) 783-3238. Cost is $3.25. Ask for stock number 055-000-00281-9.

"Compost Solutions to Dockside Fish Wastes" describes how to turn fish wastes into useful compost and fertilizer. Sea Grant Institute, University of Wisconsin-Madison, 1800 University Ave., Madison, Wisc. 53705.

Gray Water Use in the Landscape, by Robert Kourik (1988), Metamorphic Press, Box 1841, Santa Rosa, Calif. 95402. Discusses using "gray water" from bathtubs, sinks and showers for outdoor watering.

Landscaping for Water Conservation. Publications Center, Cook College, Rutgers University, New Brunswick, N.J. 08903.

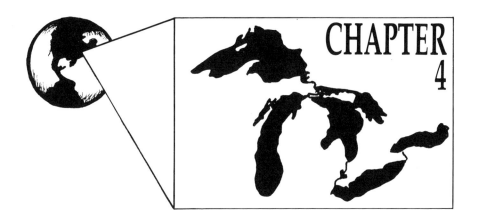

Our Environment Is Getting Trashed

offee grinds, half of a liverwurst sandwich, Q-tips, phone books, crumpled Kleenex, beer cans, apple cores, newspapers, disposable diapers, peanut butter jars, milk jugs, toothpaste tubes, plastic bags, six pack rings, smelly shoes with holes in them.

Garbage. Rubbish. Trash. Refuse. Junk. Solid waste.

We can no longer afford the luxury of throwing it away and forgetting about it. But what are we going to do with it all?

Across the country, cities are panicking about how to manage the mountains of trash created daily as their landfills become perilously full. In fact, if you dumped all of the trash we create in America into garbage trucks and lined them up end to end, they would wrap around the globe—twice. Michigan is no exception to this national solid waste crisis. People who live in Michigan generate enough trash to fill the Pontiac Silverdome to its rim in a day and a half. Each person discards an average of more than a ton of garbage each year.

Currently, most of this garbage is buried in landfills. Michigan has about 70 licensed municipal waste landfills and three or four illegal, unlicensed facilities which need to be closed, according to the Department of Natural Resources. Within five to 10 years, most of the licensed facilities will be filled to capacity and closed.

In the old days, landfills were better known as "dumps." Every township, village and city had its own small community dump—usually located on land that was considered of no value for any other use. Often this meant the dump was situated on a marsh or wetland, where contaminants would quickly seep into the ground because of the site's high water table. Garbage was piled up and burned each day, creating particulate air pollution and attracting rats and bugs that spread disease. The state first attempted to regulate solid waste disposal in 1965 by establishing a landfill licensing system, banning trash fires and requiring that dumps be covered routinely with dirt.

It was not until 1973 that the state legislature began to address the problem of groundwater contamination by requiring that new landfills only be built in areas with relatively impermeable soil. Because much of the state has a porous soil composition, this restriction dramatically reduced the number of new landfills that could be built in Michigan. Currently, landfills that accept municipal garbage (also called type II or sanitary landfills) are regulated under the 1978 Waste Management Act. The legislation set more stringent standards for environmentally-sound landfill design than any previous legislation and is enforced by the DNR.

For example, the law says that all new landfills must be equipped with a thick plastic liner and a three-foot, non-porous clay liner designed to keep waste isolated from the soil. Landfill owners also must install a series of monitoring wells to alert operators if any liquid, or leachate, has escaped the protective liner. Collection systems remove leachate before it can contaminate groundwater.

A sanitary landfill works like this: Each day garbage trucks haul mounds of compressed trash to the landfill. The daily deposit of garbage is compacted into a specific section of the landfill, called a cell, with a bulldozer. It is then covered with a layer of dirt or plastic to keep out rodents and water. The landfill owners continue to add more and more garbage until they run out of space and have to close down.

Most of the state's unlicensed landfills were built on sandy areas in the north and were not constructed with protective plastic liners. After years of continuous use, these facilities are beginning to leak. Often, they are filled with rotting organic waste, corroded car batteries, used oil and other poisonous household substances that form a toxic soup which can seep into the ground and pollute drinking water. Because the potential for groundwater contamination is so great, the state is forcing these communities to either make some expensive technological improvements or close their landfills and ship the waste to a safer spot.

The Michigan Department of Natural Resources estimates that landfills are responsible for 13 percent of known groundwater contamination in Michigan and 47 percent of suspected contamination. The costs for cleaning up contaminated groundwater are steep. For example, it cost $10 million to clean up the polluted water underneath the closed 40-acre Gratiot County landfill.

But acceptable sites for new landfills are hard to come by because anyone who lives near it strenuously objects to the potential litter, odor and water problems and lowered property values. A recent trend has been toward fewer, larger landfill operations that are waste refuges for many communities, and in some cases, several states. For example, as the number of acceptable landfill sites diminishes in the East, more garbage is being shipped to the Midwest. However, Michigan law dictates that a county can accept out-of-county waste only if permission is granted in the county plan.

In 1988, the Michigan Natural Resources Commission, the panel that oversees the DNR, adopted a Solid Waste Policy which outlines its plan for solving the state's solid waste predicament. A top priority is to reduce our dependency on landfills.Because each community has different waste disposal needs, the policy calls for each county to make its own solid waste management plan for the next 20 years, to be reviewed and updated every five years.

Clearly, landfills are not the best solution to our solid waste problem. But what are the alternatives? Individuals create waste. As an individual, you can help by cutting your personal contribution to Michigan's waste stream. We are all part of the problem—and part of the solution. If all nine million people who live in Michigan try to be more responsible about solid waste generation and disposal, our mountainous garbage problem will lessen.

The first step toward kicking the landfill habit is to create less waste by consuming less. Don't buy six apples packaged on a foam tray, wrapped in plastic wrap, then placed into a plastic bag. Instead, go to a farmers' market, pick out your own apples and put them into your reusable cloth bag. You'll eliminate three layers of unnecessary packaging that would take up space in a landfill indefinitely. You'll also save money by not paying for the multi-level packaging.

Packaging grew along with our country's desire to save time. In fact, one-third of all of our garbage is packaging material. Consumers have demanded single serving frozen dinners and desserts and have been willing to pay top dollar for convenience items such as triple- and quadruple-wrapped microwave dinners, squeezable ketchup bottles made with six different kinds of plastic and throwaway cameras. This disposable mentality is gradually giving way to greater awareness of the environment at the checkout stand. People are beginning to be equally concerned about a product's impact on the earth as well as the effect it will have on their hearts or cholesterol levels.

A recent survey of 1,000 people found that more than half passed over a product because of its excess packaging and its impact on the environment. And three-fourths of the people polled said they would be willing to pay as much as five percent more for a product packaged in recyclable or biodegradable materials. Sixty-one percent of the respondents said they are more inclined to give their business to a store or restaurant that practices good ecology by reducing plastics and using more recycled and recyclable products.

Most packaging is either plastic or foam. Because both materials are extremely lightweight, they take up more volume than weight in a landfill. Plastic makes up about seven percent of our waste stream by weight, but 14 percent by volume. As most people know, neither plastic nor foam will break down and disperse safely into the environment. Conversely, because they are inert, they will never degrade and produce toxic chemicals.

But plastics do cause trouble when they aren't placed in a landfill. Plastic and polystyrene blow into the water from landfills located near the shore, get dumped by thoughtless boaters, discarded by anglers or left behind by campers and sun worshipers. Most people are not aware of the gruesome consequences of their carelessness.

Birds, such as the kingfisher, that dive into the water in search of fish sometimes plunge right into a six-pack loop, which is virtually impossible to see beneath the water. The plastic stays around their necks and if they happen to fly by a log and snag their new plastic collar, they will be strangled. The same thing can happen if a bird catches some fishing line on its wings and gets stuck in a tree. It will most likely hang there until it starves to death, or until someone happens by and frees it.

In a few cases, fish have swum through a ring when young and retained it as they grow larger. The ring tightens and slowly squeezes the fish to death. Luckily, Michigan law now requires plastic rings to be made of biodegradable material.

Small pellets of plastic or foam are another source of danger because they resemble many natural sources of aquatic food. To a bird or a turtle, they may look like delectable fish eggs. After they gobble them up, the plastic or foam pieces will block their intestines. If they eat too much it will kill them. Ingesting buoyant plastic and foam also can kill a turtle because it cannot dive underneath the water to collect its food.

This cruel and unnecessary punishment of wildlife results from one simple act: littering. Plastic kills fish and birds in Michigan simply because humans have forgotten to think about the consequences of their actions. The plastic threat to wildlife can be solved easily if people use common sense and act responsibly.

Although they have only recently appeared on the market, "biodegradable plastics" were invented during the energy crunch of the 1970s when petroleum was scarce. Scientists added cornstarch to products as a substitute for plastic materials. But when the experimental plastic fell apart into tiny pieces the project was abandoned, only to be picked up again 20 years later with the recent wave of environmental concern.

While they are often called biodegradable, these new plastic products are not. The term "biodegradable" refers to a material that can be digested by microorganisms. So far, no one has been able to find any bugs that can digest plastic. Plastic is actually photodegradable, which means it will only break down in the presence of sunlight. Buried under several tons of garbage in a landfill, these so-called degradable plastic bags are no more likely to disappear than their regular plastic cousins.

If they do manage to make their way into a patch of sunlight, they will only fall apart into smaller (but indestructible) pieces of plastic and remain on the earth for many years. Even manufacturers of degradable plastic are not sure what will happen to their product 50 years from now.

Degradable plastics hurt efforts to recycle plastic because if they are mixed into a batch of real plastic that is melted down and reprocessed, they can weaken the composition of the benches and tables that are currently made from recycled plastic materials. Degradable plastics also use up more raw materials because manufacturers have to make them thicker to compensate for the weakening effect of the cornstarch. People who want to help the environment are being misled by degradable plastics. If you must choose plastic over paper, choose regular plastic bags and recycle them instead of throwing another bag into the landfill or, worse, littering, thinking the sun will supposedly get rid of it.

Photodegradable six-pack rings are potentially beneficial because if they degrade fast enough, they could help ease plastic's deadly threat to wildlife. A better solution would be to come up with a different way to carry our cans which doesn't involve plastic in any of its forms. Degradable plastics are simply not a viable substitute for a package made from a renewable, recyclable resource such as paper, or better yet, no packaging at all.

Compared with plastic and foam, paper seems to be a much more ecologically sound alternative. It is biodegradable and made from a renewable resource. While this is true, some paper companies cut down trees faster than nature can replace them. And while it will break down naturally under the right conditions, most of the paper sitting in our landfills hasn't broken down.

This is because landfill operators try to keep as much water as possible out of the facility to avoid groundwater contamination. A modern sanitary landfill is designed to isolate wastes and liquid from the environment. But without the presence of water as a pathway for microorganisms, garbage takes much longer to degrade. A clearly legible newspaper from 1952 was recently found in an excavated landfill in Arizona.

Paper represents a whopping 40 percent of garbage by weight and volume. Newspapers alone comprise 14 percent of landfill volume and only 10 percent of all newsprint is recycled. Only about 10 percent of the nation's newspapers are printed on recycled paper and some newspaper publishers have landfilled papers that didn't sell instead of recycling them. Phone books are another paper culprit taking up space in landfills. Each ton of paper that is recycled saves about 17 trees, 25 barrels of oil, 7,000 gallons of water and three cubic yards of landfill space. It also prevents 60 pounds of air pollution.

While plastics and foam create hazardous wastes when they are manufactured and incinerated, the tremendous amounts of energy used in the paper production process creates sulfur emissions that contribute to acid rain. The production of bleached paper products like white paper bags, paper towel, facial tissue and toilet paper creates its own danger. A by-product of the bleaching process is the carcinogen dioxin which is released during production and when paper is burned.

If people in Michigan reuse more items, they could prevent 1,600 tons of garbage from going to landfills, which is about five percent of the solid waste we generate in Michigan every day. Many items on the market are available in both disposable and durable forms. For example, choose metal razors over disposable plastic, ceramic dishes over paper plates and cloth diapers over disposable plastic ones. Favoring the reusable product is a simple, yet highly effective way to create less waste. Reusing materials sparks a creative spark in many people. They can turn egg cartons, old clothes, scraps of paper and gift wrap into practical items or works of art.

Composting can prevent more waste from unnecessarily clogging landfills and can produce fertilizer to beautify yards and parks. A compost pile duplicates and speeds up natural decomposition. The process can turn weeds, grass clippings, leaves and food wastes into a nutrient-rich fertilizer. It can protect gardens from weeds and improve the soil's moisture retention.

Composting is catching on. Before the Legislature mandated yard waste composting, less than one percent of yard wastes were composted, even though it comprised 20 percent of the solid waste stream and as much as 75 percent in the fall. Communities have found that compost saves money because they don't have to purchase chemical fertilizers and it is a safe alternative to air-polluting leaf burning. State law bans landfilling or burning yard waste beginning in 1994.

For decades, we have been throwing away millions of dollars in energy, raw materials and natural resources. No more with recycling. Instead, products made from recyclable materials such as glass, paper, metal, plastic, oil and foam can be reprocessed and shaped into new products, saving energy and resources. Michigan residents recycle about 12 percent of the waste stream, but the state Solid Waste Plan wants them to boost that to 30 percent. Some experts say as much as 65 percent of the nation's waste is recyclable. Others are more cautious and say only about 45 percent can be recycled and that amount depends heavily on the definition of "waste stream."

Many counties in Michigan now have recycling facilities where residents can drop off recyclable items. Larger urban areas such as Lansing, Detroit, Grand Rapids and Kalamazoo have curb-side service, where recycled materials are picked up along residential streets once or twice a month. And cities have been very ingenious in encouraging recycling. In Grand Rapids recycled goods are placed at the curb in specially marked clear plastic bags. Mixed trash goes into regular bags. Here's the catch: Regular garbage bags cost 35 cents and clear ones used for recycled materials cost 10 cents.

The state has gotten into the act, too. It offers money to communities that need help in setting up drop-off services, curb-side collection, office paper recycling, collection at apartment complexes, used oil and battery recycling and household hazardous waste collections. The DNR Solid Waste Management Division has $29 million from the "Clean Michigan Fund" and $150 million through an $800 million bond program to lend or give to communities for solid waste projects.

Communities starting a recycling program, helping recyclers purchase equipment or helping small companies use recycled materials in their products can qualify for state grants. Funding also is available to educate people that recycled products are not inferior and to develop markets for recycled materials.

The most dramatic illustration of the positive impact recycling has had on our solid waste problem is the success of the Michigan bottle bill, which took effect in 1978. According to DNR studies, the bottle bill, which was brought about through the efforts of Michigan United Conservation Clubs, has reduced the state's waste stream by about eight percent. So far, the bill has kept six million tons of refuse out of municipal landfills, saving taxpayers an average of $37.5 million in disposal costs during 1988 alone. And because millions of tons of cans and bottles are being returned to stores to be recycled, the bill has reduced roadside bottle and can litter by 90 percent. In fact, Michigan has the highest container return rate of any of the 10 states that have beverage container laws—about 93 percent. The remaining seven percent are either thrown out, lost, taken out of state or broken.

Recycling cans and reusing glass bottles conserves the energy resources that would have been used to make a brand new container. For instance, a recycled aluminum can represents a 95 percent energy savings over those that are made from raw materials. The increased recycling due to the bottle bill has created nearly 5,000 jobs.

And the law keeps getting better. In 1990, the Legislature diverted 75 percent of the $380 million worth of unclaimed deposits to a conservation trust fund. After the fund has built up for 10 years, the interest from the $33 million in annual revenue will be used for solid waste cleanup and prevention. However, bottlers are challenging the law in court.

But recycling only works if there are markets for the recycled materials. More industries must put recycled materials to new uses because if no one buys materials from recycling centers, those centers will go out of business.

While it is extremely important for people to reduce the waste stream through recycling, it also is essential that we give our business to companies that use recycled materials to make their products. Most recycled products are printed with a slogan that says "printed on recycled paper," "made with recycled materials" or "recycled."

Scores of companies already use recycled materials. Dart Container Corporation in Mason, Mich., recycles packaging foam and foam cartons and cups into hard plastic pellets, which can be turned into kitchen utensils, plastic car parts and paint brush handles.

Another new arrival on the Michigan recycling scene is Processed Plastics of Ionia. It melts down milk jugs, soap bottles and other household plastic containers collected by recycling centers around the state and turns them into plastic lumber. In some cases, plastic lumber works better than real wood because it is water resistant. It can be used for boat docks, picnic tables and outdoor benches. Farmers have found that the alternative lumber works well for animal feeders and pens.

Other uses for recycled plastic include base cups for soft drink bottles, flower pots, fiberfill for ski jackets and sleeping bags, home insulation and toys.

However, one major market for recycled foam, plastic and paper—the food service industry—has been closed off by the U.S. Food and Drug Administration and state sanitary laws that require that food packaging be made from raw materials because they believe recycled containers pose a possible health threat. Therefore, schools and hospitals cannot close the recycling loop by purchasing recycled cups and plates after recycling the items they use. It is a huge void because the food industry is the nation's largest user of packaging, consuming almost 45 percent of all of the packaging that is produced.

Scrap tires are another huge portion of solid waste that can be recycled. Nearly eight million worn tires are discarded each year in Michigan. They sit heaped in huge piles on farms and in gravel pits across the state, creating the perfect breeding ground for mosquitoes as they fill with water and bake in the sun. Unregulated tire heaps also pose a tremendous fire hazard because once a tire ignites, it is nearly impossible to extinguish. The blazing tire melts, spewing hazardous chemicals into the air and contaminating groundwater or nearby surface water.

Other tires find their way into municipal landfills where they hog space and cause problems when they fill with air and pop up to the surface, sometimes at high speeds. Recent legislation regulates used tire collection and storage in Michigan in order to reduce the health and fire risks associated with the piles.

There is a simple way to reduce the number of used tires clogging up our landfills—use retreaded tires. The product, which costs 60 percent less than a new tire, has a new coating of rubber treads over the old tire casings. The option saves oil, too. A car tire requires about seven gallons of oil to manufacture, while retreading a tire only uses 2.5 gallons. Retreading a truck tire saves about 15 gallons of oil.

Retreaded tires already are widely used on ambulances, school buses, trucks, race cars and airplanes. Scrap tires also can be shredded and reprocessed to supplement or substitute rubber in mats, kitchen accessories and garbage cans. In New Mexico, people build houses out of them.

The recycled steel and scrap-metal business always has done well in Michigan because the automobile industry creates a strong market for the materials. The Kellogg Company in Battle Creek uses recycled paperboard or grayboard (named for its color) for cereal boxes.

But it still is cheaper for many companies to use raw materials. To change that, the state may have to offer tax incentives, financial assistance or loans to companies that use recycled materials.

As more and more local communities incorporate curb-side or mandatory recycling programs, state and local governments must ensure that the demand for recycled goods increases at the same rate as the supply of recycled paper, glass, plastic and metal. State and local governments can create an instant market for recycled products by buying only recycled paper products, using retreaded tires for their vehicles and installing benches, picnic tables and signs made from recycled plastic lumber. It is hoped such actions will inspire citizens and businesses to follow suit.

On another front, the by-products of landfills can serve useful purposes. A closed landfill produces methane as the enclosed garbage begins to decay. This can be turned into a fuel source for local homes or industries. Landfill cells begin producing methane within three to five years. Within 10 years most of Michigan's landfills will be closed, so we may have an energy source for 40 years or more.

Here's how the process works. When food wastes and other trash are sealed in a landfill under anaerobic conditions (in the absence of oxygen-rich air) they are consumed by bacteria and converted into methane, carbon dioxide and ethanol, or alcohol. Landfill operators collect the methane gas in a set of wells, or pipes, which are drilled about 40 feet into the pile of trash. A vacuum sucks the gas from the trash into a pipeline which carries it to equipment that cools and compresses the gas to rid it of moisture. The methane gas then can be transported to a power plant to be burned for heat, air conditioning or electricity.

In areas with extremely large landfills, methane gas provides electricity for many neighborhoods. One methane power plant in Los Angeles generates enough electricity for nearly 200,000 homes. Michigan's first landfill gas project was constructed in 1985 by Granger Waste Management Company at its Lansing Township landfill. The methane is piped about a mile away to Motor Wheel Corporation, which uses it as boiler fuel. The facility provides 525,000 cubic feet of gas a day, saving about 630,000 gallons of crude oil every year.

Another successful methane energy project is under way at the Wayne Disposal landfill near Ypsilanti. That company sells its methane gas to Detroit Edison, which converts it into electricity for approximately 2,000 homes. One-quarter of the energy that the landfill generates is used on-site to heat and power Wayne Disposal's headquarter offices. The methane also heats an experimental "hydroponic" greenhouse perched atop a closed cell in which gourmet varieties of lettuce are grown in nutrient-enriched water—with no soil.

There are about 150 landfills in Michigan that could supply enough gas for electricity for anywhere from 1,000 to 5,000 homes, according to estimates by landfill gas experts. Within five to 10 years that number will grow as landfills are sealed and begin producing gas. Because less water can seep into newer landfills, the garbage inside will take longer to decompose and the gas supply will last longer.

But an overabundance of energy deters utilities from buying methane gas as an energy source. State law requires utility companies to buy energy from small waste-to-energy suppliers, but the utilities won't pay enough to cover the costs of providing methane gas. This is because the state's two major utilities, Consumers Power Company and Detroit Edison, have a surplus of energy supplies and have recently built power facilities in Monroe and Midland. They would rather use those energy sources than negotiate with small energy producers.

And methane, while it lasts, is a clean fuel. Burning methane for power instead of coal reduces emissions of sulfur, which leads to acid rain, and CO_2, which contributes to global warming. But politics and economics are working against such long-range thinking in favor of short-term gains. Many landfill operators have been forced to flare off methane to prevent explosions because no one will buy it at a reasonable rate.

The next challenge is to figure out what to do with all of the land that closed landfills occupy. Wayne Disposal wants to plant a Christmas tree farm atop its closed facility and in Lyon Township north of Ann Arbor plans were unveiled to build a large recreation park on its old landfill. The park will include baseball diamonds, tennis courts, paths and sledding hills. Brighton and Riverview boast of landfills-turned-ski hills.

As more landfills begin to close around the state, communities can take advantage of the opportunity to "make good" on land that has typically been perceived as useless for anything other than garbage dumps. Although closed landfills are not suitable building sites, they can house amphitheaters, golf courses or recreation areas.

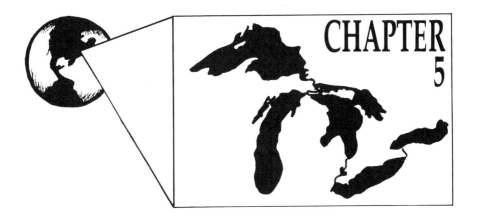

The Hazards of Incineration

A more controversial garbage disposal system is incineration. Like methane gas recovery from landfills, incinerators can turn wastes into elecricity through the combustion process. Instead of burning coal to produce electricity, an incinerator burns garbage. Truckloads of trash are fed into a furnace where it falls onto a moving metal grate and is exposed to intense temperatures of 1,800 degrees to 2,400 degrees Fahrenheit. The burning garbage boils water, creating steam that drives a turbine. The turbine produces electricity which can be sold to a utility company for profit.

There are thousands of incinerators in Michigan, from one building burners to Detroit's huge municipal incinerator. Incinerators burn about 20 percent of the state's solid waste stream and the 1988 Michigan Solid Waste Policy sets a goal of increasing that to 40 percent by 2000. But with increasing incidents of hazardous air emissions, that percentage may be re-evaluated until pollution control equipment becomes more affordable.

The existing waste-to-energy plants were built in heavily populated areas with high volumes of waste—Detroit, Grand Rapids and Jackson, as well as many other areas. Incinerators' expense prohibits smaller communities from considering them. The Detroit plant, the largest in the state, cost $438 million.

The benefit of incineration is that it reduces the bulk of garbage by 90 percent, leaving only ash to be disposed. That poses a problem because any chemicals that were present in the original garbage will be present in the ash, but in a more concentrated and toxic form. Among the most dangerous components of incinerator ash are the heavy metals of lead, nickel and cadmium. Nickel and cadmium come from household batteries.

The incinerator's high temperatures turn liquids and solids into dangerous gases. Dioxins and furans, which can cause miscarriages and birth defects, are frequently found in incinerator ash and emissions when bleached paper and some plastics are burned. Mercury fumes also are created when automobile batteries are burned. Although emissions of hazardous particles and gas are regulated by the Michigan Clean Air Act, testing of incinerator ash for hazardous materials was first done only recently.

Now that the ash has been found to be toxic, it must be transported to a costly monofill. Before testing, it was sent to municipal landfills. Standards for monofills are in between those for sanitary landfills and those for hazardous waste. They must keep groundwater and surface water out of the cell and free of contamination through a system of isolating liners, leak detection and collection systems. The facility also must be covered to prevent hazardous ash from becoming airborne.

The law now requires that new waste-to-energy incinerators use the best available pollution control technology to trap hazardous emissions before they leave the smokestack. The most effective air pollution equipment for incinerators on the market today is a system that combines an acid gas scrubber and fabric bag filter. The EPA has said that when combined with proper combustion techniques, acid scrubbers and particulate filters can reduce hazardous emissions by as much as 99 percent.

When sent through an acid gas scrubber and bag filter, the acid gases and particulate matter generated by the burning waste are blown into a dry scrubber where they are sprayed with alkaline lime to neutralize the acids. Then they are put into a bag filter, which works like a vacuum cleaner. When the particles are sucked through the filter they get caught on the fabric sides of the bag, leaving primarily clean air to escape out the smokestack.

Acid gas scrubbers and bag filters are extremely expensive. For that reason, most of the incinerators in the state are equipped with electrostatic precipitators which are not as effective in reducing hazardous air emissions. These devices magnetize loose fly ash and particulates with an electric charge, causing the particles to attach themselves to a neutral steel plate. Plant operators can shake the particles loose from the plate into a hopper below and dispose of them. While electrostatic precipitators are quite effective at controlling fly ash, they still allow harmful hydrogen chloride, sulfur dioxide, mercury and lead emissions to escape.

A simpler and more economical way to reduce hazardous ash and emissions is to remove from the trash the components that create the problem in the first place. Recycling, composting and source reduction before incinerating results in less hazardous ash and air pollution. Recycling is more important than ever in communities with incinerators.

And consumers are mostly to blame for the flow of materials that emit toxic elements. Hazardous gases are formed when the following items are incinerated: automobile oil, brake fluid and antifreeze; car, flashlight and watch batteries; home cleaning and gardening products; and paints and solvents. When these gases combine, they react, producing extremely dangerous airborne emissions. Plastic and bleached paper packaging also release hazardous fumes when burned in an incinerator.

Recycling and collection programs for plastics, household hazardous wastes, used oil and batteries should be made mandatory in communities considering an incinerator. However, the state has allowed municipal garbage to be indiscriminately burned, no matter how many hazardous materials it may contain. As a result, facilities like the Detroit incinerator have been exposing local residents to unsafe levels of mercury, lead and other hazardous emissions.

Separated loads of trash require smaller incinerator capacity. Smaller incinerators do not have as many problems as larger ones like Detroit, which can burn about 3,000 tons of garbage per day. Removing glass and wet items allows more heat to be generated when trash is burned, reducing the need to purchase extra fuel to keep the temperatures high enough. The increased heat generated by drier garbage can be converted to greater amounts of energy.

Because incinerators can reduce the volume of municipal solid waste by 80 percent to 90 percent, they have become an attractive alternative as many landfills fill up. But they are risky. Former DNR Director David Hales said Michigan will not invest in any more incinerators until cleaner, more efficient technology is available. Even if incinerator operators are eventually able to perfect the combustion process and install the best pollution control equipment in order to increase safety, the costs of building and running the plant are so prohibitive that many citizens may wonder if it is worth it. Plus, waste-to-energy plants quickly become less effective as they age.

Of all the options available to minimize our solid waste problem the two most important solutions—source reduction and recycling—also are the simplest, least expensive and the kindest to the environment. These are common-sense alternatives. Had we followed through with them several years ago, we would not find ourselves up to our necks in garbage today. But it's never too late to start a good thing.

Landfills and incinerators will be more environmentally sound if recycling, source reduction and reuse are widely adopted. The technology exists to build safe landfills and incinerators but they are overburdened with excessive amounts of garbage that contain hazardous chemicals. Once these are removed through recycling and reduction, those alternatives will work quite well.

The solid waste dilemma is one of the more manageable environmental challenges our state will have to tackle over the next few years. The amount of trash we accumulate in Michigan can be dramatically reduced if people, businesses and local governments are brave enough to alter slightly their usual patterns of behavior for the good of all.

SOLUTIONS
Things to do:

• Do not buy anything with many layers of wrapping that is neither biodegradable nor recyclable.

• Buy food in bulk and at farm markets and roadside stands. It is cheaper and eliminates unnecessary bags and containers.

• Bring your own fabric or net bag to the grocery store. If you forget it, ask for paper because it is biodegradable and is made from a renewable resource. If you have to take plastic, reuse it or recycle it. Several grocery stores reimburse customers when they return paper and plastic bags.

• Purchase eggs in cardboard cartons.

• Buy products made from recycled materials. This creates a market for the items that people separate from their trash for recycling.

• Buy stationary, envelopes and office paper that is made from recycled fibers. Use recycled paper for copies and printing and encourage its use at your workplace. Write on both sides of the paper and save scrap paper to use again.

• Shop at places that carry re-refined motor oil and collect used oil for recycling.

• Buy unbleached paper products if possible.

• Look for cleaning products in refillable containers.

• Avoid disposable plates, cups, bowls and utensils.

• Use cloth rags instead of paper towels.

• Steer clear of disposable, throwaway products. Use a refillable fountain pen instead of a disposable ball-point pen. Try a reusable metal razor instead of a plastic one.

• Use a cloth handkerchief instead of tissues.

• Reuse plastic bags, foil, plastic wrap, glass or plastic containers and squeeze bottles.

- Use cloth diapers. Americans throw out four million tons of disposable diapers a year because each child goes through about 2,500 diapers a year. A disposable diaper takes 500 years to decompose, a cloth one takes one to six months.

- A ceramic coffee mug is far preferable to paper or foam cups. Encourage others to bring a mug to work. Carry a lunch box instead of paper bags.

- Stop junk mail! Americans receive nearly two million tons of junk mail each year and the average person spends eight months of his or her life just opening it. To stop unwanted mail, write to Mail Preference Service, Direct Marketing Association, 6 East 43rd St., New York, N.Y. 10017.

- Make your possessions last. Mend and repair broken or worn items and buy items that were meant to endure.

- Donate unwanted items to a local charity or have a garage sale. Give to a library old books, magazines and newspapers or recycle them.

- Ride on retreaded tires.

- Build outdoor decks, docks and patios with plastic lumber. It lasts longer than wood when used outside.

Things to do in your community:

- Insist on a strong, effective recycling program in your community.

- Support a ban on packaging materials that are not recyclable or biodegradable.

- Start a recycling program at your school, workplace or apartment building.

- Volunteer at your local recycling center or make a donation to help keep it operating.

- Encourage your utility company to buy methane-powered electricity from a local landfill. Work to turn filled landfills into a park, ski area or wildlife habitat.

Contacts:

• Michigan Department of Natural Resources, Resource Recovery Section, Box 30241, Lansing, Mich. 48909, (517) 373-0540. It can provide information on composting and recycling programs, as well as the Michigan Recycling Directory.

• Michigan Recycling Coalition, Box 10240, Lansing, Mich. 48901.

• Dart Container Corporation in Mason, Mich., accepts foam plates, cups, trays and packing materials for recycling. Recyclers can drop off foam at 432 Hogsback Road in Mason, or call (517) 676-3800.

• U.S. Environmental Protection Agency Waste Minimization Hot Line: 1-800-424-9346

• Contact Granger Waste Management Company for a free brochure about converting methane gas from landfills into energy, 3535 Wood Road, Lansing, Mich. 48906, (517) 372-2800.

• For a free catalog of recycled paper products, write to: Earth Care Paper Inc., Box 3335, Madison, Wisc. 53704, or call (608) 256-5522; French Paper Company, Box 398, Niles, Mich. 49120, (616) 683-1100; and Peninsula Paper Company, 1000 North Huron St., Ypsilanti, Mich. 48917, (313) 482-2600.

• Companies that use shredded newsprint products: Applegate Insulation Systems Inc., Box 292, Okemos, Mich. 48864, (517) 349-0466. Total Comfort Insulation Corporation, 85 Minnesota, Troy, Mich. 48083, (313) 588-4232; and U.S. Cellulose Corporation, 5020 West River Drive, Comstock Park, Mich. 49321, (616) 784-6447.

• To find out more about plastic lumber products, contact Processed Plastics Company, Box 69, Ionia, Mich. 48846, (616) 527-6677.

• For information about retreaded tires, contact the Michigan Retailers Association, 221 North Pine St., Lansing, Mich. 48933, (517) 372-5656.

• To find out where to buy re-refined automobile oil, contact West Michigan Environmental Action Council, 1324 Lake Drive SE, Grand Rapids, Mich. 49506, (800) 634-9504; or Michigan State University Cooperative Extension Service, MSU, East Lansing, Mich. 48824-1222, (517) 355-2308.

• The American Petroleum Institute's "Recycle Used Motor Oil: A Model Program" helps people set up an oil recycling program in their communities. 1220 L Street NW, Washington, D.C. 20005.

Further reading:

Coming Full Circle, a book by the Environmental Defense Fund that explains how to form a local recycling program. Cost is $10. Write: Environmental Defense Fund, 1616 P St. NW, Washington, D.C. 20036.

Keep Earth Clean, Blue and Green: Environmental Activities for Young People, by George and Dorothy Hemmings. Citation Press, New York, NY, 1976.

Recyclopedia: Games, Science Equipment and Crafts from Recycled Materials by Robin Simons, developed at the Boston Children's Museum. Houghton Mifflin Company, Boston, 1976.

"There Lived a Wicked Dragon," U.S. Environmental Protection Agency, 1973. An environmental coloring book for children and adults.

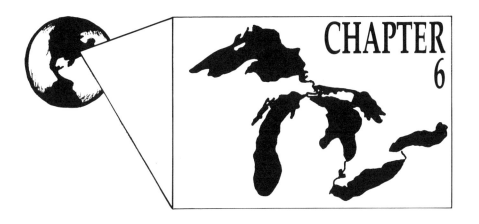

Radioactive Waste, Toxins and Heavy Metals

Under the Midwest Compact, Michigan has been chosen as the 20-year repository for all of the low-level radioactive waste generated by six other states. This coalition of Michigan, Ohio, Indiana, Minnesota, Wisconsin, Iowa and Missouri was formed under the federal Low-Level Radioactive Waste Policy Act of 1980, which forces states by 1993 to either form compacts or decide to take care of their own waste. Compact states are eligible for federal funds, generated by a surcharge on waste produced, to pay for a disposal site and to clean up contamination. Solo states are responsible for all costs and cannot refuse waste from other states.

Several states, including Michigan, have sued the federal government saying that the law intrudes on the states' sovereignty by forcing them to become responsible for someone else's waste.

Michigan was chosen as the "host" state because it generates the most radioactive waste in the region. The responsibility rotates every 20 years, with Indiana up next. But debate rages over whether Michigan should be chosen as a site because of its high water table and proximity to the country's largest supply of fresh water—the Great Lakes. Environmentally, the dump should be in a flat place, with low population density, impermeable soil (thus less potential for groundwater infiltration) and minimal impact on wildlife or endangered species. Once such a site is found, the facility should stay there for good to avoid a proliferation of sites.

Gov. John Engler wants Michigan to pull out of the compact and work to change the federal law to exempt Michigan from having to build a site. Michigan has since been thrown out of the compact.

What is low-level radioactive waste? Fifty to 75 percent of its volume is generated by nuclear power plants, consisting of contaminated protective clothing, instruments or valves. The rest comes from industry, government, hospitals and universities.

Three different classifications have been assigned to different types of low-level radioactive waste. Ninety-seven percent of it is considered type A and B, which means it will take up to 100 years to decay and become harmless. The remaining 3 percent, type C, will stay "hot" for 500 years. Most of the hospital and lab wastes are type A and B, while some material from nuclear power plants is type C.

Before Michigan was thrown out of the compact for allegedly being uncooperative, it had narrowed the list to 78 sites but faced fierce community oppostion across the state. When—or if—a site is selected, the disposal facility will be built underneath the ground with two-foot-thick concrete walls. Radioactive wastes will be placed into concrete containers that can hold 10 metal drums. Monitoring equipment and collection pipes will be installed underneath the containers to capture possible leachate and the 20-acre facility will be surrounded by a 2,500-acre buffer zone. Experts believe they have designed a safe facility—at least for the short term—but no disposal method has proven to be leak free over several decades.

While Michigan contemplates radioactive waste disposal, a host of chemicals have already made themselves at home in Michigan.

Dioxins, furans, polychlorinated biphenyls, DDT, PBB, lead, methylmercury, benzene, arsenic, cadmium. These malevolent-sounding chemicals are scary. Scientists say that because so many hazardous substances contaminate our ground and water, people who live in Michigan and the Great Lakes basin are exposed to higher levels of dangerous chemicals than residents in other parts of the country. Because hazardous pollutants are everywhere in our ecosystem threatening us and wildlife, we have a responsibility to learn about their arrival, risks, and—most importantly—their removal.

Cadmium

Cadmium is a heavy metal that is picked up by tobacco plants from phosphate fertilizers. The substance ends up in cigarettes when they are manufactured. Thus smokers—and people who are frequently exposed to cigarette smoke—have higher levels of cadmium in their bodies. It gradually accumulates in the liver and kidneys and can cause high blood pressure and hardening of the arteries, which leads to heart disease. Cadmium is released when batteries and coal is burned.

Lead

The amount of lead we are exposed to has been greatly reduced since lead was phased out of gasoline. But some lead is produced in diesel fuel combustion and it arises when coal is burned in power plants. Lead used in plumbing solder and to seal tin cans can result in food and water contamination. Lead plumbing solder was banned in 1987 for use in public water systems, although it still exists.

Inhaled particles of lead paint also are dangerous. Approximately 40 million U.S. homes built before 1950 and 20 percent of the homes built between 1960 and 1975 are potential sources of lead contamination because of the lead-laden paint used to cover them. It was not until 1976 that the legal paint lead levels were lowered. Children are particularly at risk for lead poisoning because their bodies absorb it more quickly than adults. About 200 children die every year after inhaling lead dust from cracking paint and sucking on paint chips. Others suffer from paralysis, blindness, hearing damage and mental retardation because of lead poisoning.

Mercury

Mercury, an element found along the ocean bottom, is released into the air when coal or batteries are burned and into the water when factories dump their wastes. Incinerators alone release nearly 6,000 pounds of mercury into the atmosphere every year. Dental workers who spend a lot of time making fillings are often exposed to high levels of mercury if they are working in a poorly ventilated office. They can experience central nervous system problems such as anxiety and depression, loss of coordination and sensory function.

In its most toxic form, methyl mercury, the metal can circulate through the body and attack the brain, kidneys and liver, causing neurological problems and birth defects. The formation of this highly toxic compound is accelerated when regular mercury is mixed with acidic lake water (as we have in Michigan due to acid rain) and is converted to methyl mercury by microorganisms in the sediments. People who frequently eat fish caught in these waters have higher levels of mercury in their bodies than those who do not.

Dioxins

High levels of dioxins have been found in rivers and fish in Michigan. Dioxins and furans are a family of chlorinated hydrocarbons formed as by-products of herbicide, paper and PCP manufacturing and during the incineration of fossil fuels, plastics and bleached paper products. Dioxin was the reason homes were evacuated near Love Canal in Niagara Falls.

Much of the dioxin that contaminates the environment originates from leaking or abandoned industrial waste dumps. In its most poisonous form, TCCD, mere traces can cause liver cancer, birth defects and death in laboratory animals. Fortunately, dioxins move slowly through the soil, making cleanup efforts much more successful than other substances.

PCBs

Like dioxins and furans, polychlorinated biphenyl (PCBs) are mixtures of many different chlorinated hydrocarbon compounds. They have not been manufactured since 1979, but are still used as insulating fluids in electrical transformers. Before 1979 they were widely used to manufacture plastics, rubber, carbonless copy paper, paints, sealants and inks. They also were used as a vehicle to spread pesticides and to control dust on roads.

Dioxins and PCBs do not break down in water, but dissolve in the fatty tissue of animals and humans. Thus, they increase in concentration as they progress up the food chain. PCBs have caused reproductive problems, tumors, liver and kidney damage and skin lesions in laboratory animals. In 1980, the EPA required that all materials containing PCBs be placed in a landfill or incinerated at extremely high temperatures. But much of the PCBs that were landfilled evaporated and contaminated the atmosphere.

PCBs contained in electrical transformers in office buildings, malls and subway stations have caused serious health and environmental problems when they caught on fire. When they burn, PCBs generate even more hazardous by-products that spread into drinking water supplies and contaminate surface water. By 1985 the EPA banned the use of PCBs in buildings and enclosed areas such as subways. That leaves some 375,000 tons of PCBs in about 140,000 sealed transformers. More PCBs are released every year when those transformers accidentally leak, explode or ignite.

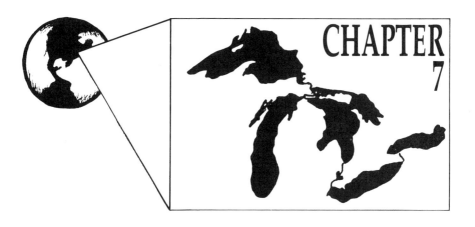

Agricultural Chemicals May Be Out Of Hand

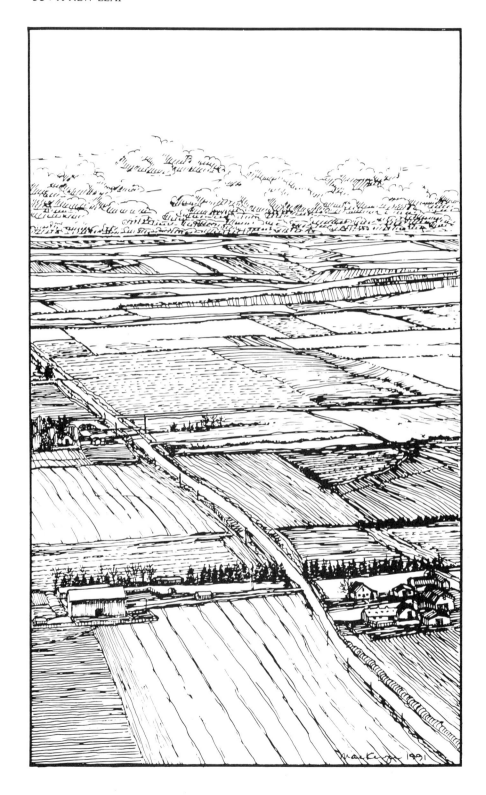

When Michigan's population began to swell in the 19th century, human activity shifted the ecosystem from a complex web of plants and animals living in a variety of habitats to a simple agricultural system. This lack of biological diversity gave rise to a swarm of uncontrollable, crop-threatening pests normally kept in check by a balanced habitat. Then the world discovered pesticides. Use of it mushroomed after World War II when DDT was discovered to be a highly effective insect killer.

DDT was widely used because it does not dissolve in water, allowing it to stick to plants even during irrigation and rain. But this characteristic made cleanup difficult once its dangers were discovered. Because DDT (a type of chlorinated hydrocarbon) persists in the environment for a long time, it builds up in animals and people who ingest it and has caused reproductive problems in birds. The EPA banned DDT in 1972, although it is still found squirreled away in barns and garages.

Other less persistent chemicals, organophosphates, have replaced DDT but are hazardous to humans and wildlife, too. Because these chemicals, which interfere with insects' nervous systems, are more water-soluble than DDT they must be applied to crops more often. In the end, they are as widespread as DDT.

Herbicides thin weeds that compete for space with crops. Some herbicides, such as Michigan corn farmers' favored Atrazine, kill unwanted foliage by disrupting photosynthesis. Systemic herbicides, such as 2,4-D and 2,4,5-T, stimulate growth hormones, which kills plants because they are unable to produce enough food to feed their rapidly growing tissue.

Herbicides and pesticides are a necessary evil. Without them, pests and plant disease would wipe out more than half of the world's food supply every year. The use of pesticides has increased food production and helped to curb hunger around the globe. (Although one study asserts that farmers lost one-third of their crops to pests before chemicals. With chemicals, they still lose a third of their crops to insects.) In the United States, the economic benefits of agricultural chemicals have been passed onto the consumer via lower food prices.

In addition to increasing crop yields, pesticides have reduced uninviting produce bugs, fungus or holes. Since agricultural chemicals enable farmers to produce more crops in less space, more land is available for other uses such as wildlife and recreation areas. Proponents of pesticides also argue that they are cheaper and act faster than other alternatives and that they are safe if used correctly.

Pesticides also have played a key role in preventing diseases such as malaria and bubonic plague, which are spread by mosquitoes, fleas, lice and flies. According to the World Health Organization, DDT and other insecticides have saved seven million people from death since they were first used in 1947.

But a huge flaw in pesticide use is that insects become immune to them. If a Michigan potato farmer sprays carbofuran, a carbamate insecticide, to control the Colorado potato beetle, most of the beetles will die. But the few beetles that manage to stay alive most likely have a genetic mutation that enabled them to survive the typically lethal dose of carbofuran. If these survivors breed, they will create a more resistant strain of Colorado potato beetle. The strongest of their offspring will survive another round of chemicals and create an even more resistant potato beetle population.

To keep up with the new strains of potato beetles, the farmer will have to apply more and more carbofuran until it no longer works effectively. Then he will have to keep switching to different (and most likely stronger) chemicals until he finds something that works on the new "superbugs," which would not have evolved if the farmer had avoided pesticides. Weeds develop a similar type of resistance to herbicides but at a slower rate than insects.

Michigan apple growers have experienced a similar problem with genetic resistance in a plant disease called apple scab. The apple scab fungus has gradually become resistant to the widely used fungicide benlate. When that chemical became ineffective against the disease, growers had to switch to a new one. Most instances of genetic resistance in weeds, fungi, bacteria and insects occur after repeated use of only one kind of pesticide.

There are drawbacks to another common farm chemical—fertilizers. Synthetic chemical fertilizers can cause soil to become compacted so that it repels water. When the water runs off, it carries with it particles of soil, nutrients, fertilizer and pesticides. Eventually that mixture will reach nearby lakes, rivers and streams. First, the influx of suspended soil particles makes the water cloudy. The fertilizer residues activate algae blooms, which deplete the water of oxygen, a necessity for fish. As more and more fish are killed, the aquatic ecosystem will shift to favor more pollution-resistant species of plants and animals.

The fish that survive, and the birds and humans who eat them, will ingest pesticides and fertilizers. The chemicals then collect in fatty tissue. Eating pesticide-contaminated fish has led to reproductive problems in several species of birds, including peregrine falcons and the bald eagle. The chemicals reduce birds' ability to produce eggs with strong shells, causing a sharp decline in population.

Because concentrated amounts of nitrates in synthetic fertilizer cannot entirely be absorbed by plants, the excess can destroy beneficial earthworms, which break down dead plants and keep the soil absorbent. Synthetic fertilizers also cause plants to develop a surface root system. Roots no longer dig deep to find nutrients because the food is all on the surface. With no roots to break it up, the soil becomes very compact, which leads to disease, harmful insects and water runoff. To counteract these problems, the gardener or farmer must continuously apply more and more chemicals to keep the plants growing.

Excessive use of agricultural chemicals is a major source of groundwater contamination in Michigan. Fertilizer nitrates and the herbicide Atrazine, which is used to treat 41 percent of the state's cropland, pollute wells in many farm communities. EPA health regulations cap Atrazine concentrations in drinking water at three parts per billion, and for nitrates, 10 parts per billion. The Michigan State University Institute of Water Research tested 38 wells near chemically-treated fields and found that 29 percent had detectable traces of Atrazine and 50 percent exceeded the EPA's nitrate limits.

Nitrates are especially dangerous if ingested by infants or young children. Once inside the stomach, they transform into nitrites, which rob the child's red blood cells of their ability to carry oxygen. This condition, commonly known as "blue baby disease," can cause infant brain damage or death. High levels of nitrates in drinking water can cause stomach cancer in adults. The health effects associated with long-term exposure to Atrazine are not as well understood. People who have ingested extremely high doses of the herbicide have experienced stomach cramps, diarrhea, vomiting and irritated eyes, nose, throat and lungs. Animal tests with Atrazine have shown mammary tumors in female rats.

Some farmers further contribute to groundwater pollution by spraying water-diluted pesticides and fertilizers through irrigation pipes. If the system has faulty valves, the chemical solution can be back-flushed through wells into the groundwater. Chlordane, a pesticide used to kill termites, often turns up in drinking water supplies. Tests have shown that chlordane is potentially carcinogenic and harmful to the central nervous system.

In this country, approximately seven million agricultural workers face the risk of health problems due to occupational pesticide use. Pesticides cause approximately 450,000 illnesses and at least 25 deaths each year among agricultural workers, according to the World Resources Institute. These numbers may be greatly underestimated because many cases of pesticide poisoning go unreported. Plus, it is nearly impossible to prove that exposure to agricultural chemicals caused an individual's death.

People who have been subjected to dangerous levels of pesticides suffer nausea, dizziness, blurred vision, loss of muscle control, respiratory paralysis, convulsions, comas and reproductive problems. The National Institute of Cancer says farmers who use the common herbicide 2, 4-D are twice as likely to develop a rare form of lymph cancer called non-Hodgkin's lymphoma. Herbicides 2,4-D and 2,4,5-T also have been linked to Hodgkin's disease and other types of cancer. If the pesticide doesn't cause problems, the toxic solvents used to spread the pesticide and make it stick pose a danger.

Some scientists believe that people who are exposed to hazardous agricultural chemicals from the food they eat also may be at risk. The U.S. Department of Agriculture estimates that 70 percent of all crops in the United States are treated with pesticides. About half of all imported fruits and vegetables tested on a random basis also contain detectable levels of pesticides, according to the federal Food and Drug Administration.

Infants and children are more at risk because hazardous agricultural chemicals tend to concentrate in breast milk and children tend to eat larger quantities of pesticide-covered produce than adults. Their immune systems are not developed enough to combat the harmful effects of these chemicals. In addition, the federal health standards for acceptable levels of pesticide residue in our drinking water and food are based on adult levels of tolerance. There are no separate regulations for children.

Our nation's over-reliance on synthetic fertilizers, insecticides and herbicides may turn out to be a Faustian bargain. In return for short-term profits and copious harvests, we could face some serious long-term consequences. These include soil sterility and erosion, expensive cleanup of contaminated surface and groundwater and indestructible new strains of insects and weeds.

Many people believe that without pesticides the world would experience widespread famine and disease. But there are alternatives to chemical pesticides. Organic pesticides offer a solution. Pyrethrum, a powder produced from chrysanthemums, was first used by the Chinese 2,000 years ago. Cutting back on the amount of synthetic pesticides used would reduce some of the environmental and health problems caused by the chemicals.

The most promising alternative to pesticides is integrated pest management (IPM), a farming technique which combines many different pest-control methods such as crop rotation and natural predators. Chemical pesticides are used as a last resort and in the smallest amounts possible. IPM treats each crop as a separate ecosystem whose control program requires an intimate knowledge of the soil, crop diseases and the pests that compete with crop growth.

Farmers who practice integrated pest management use data on meteorological conditions to predict pest outbreaks. They time plantings and harvests accordingly to keep pesticide use to a minimum. When small amounts of pesticides are used on crops, they are frequently rotated to prevent the development of genetic resistance. The goal is not to eliminate pests but to keep them at a reasonable level.

Although they do not produce the same yields as farmers using chemicals, they do cut the costs of an endless cycle of chemical pesticides, which can cost as much as $400 per pound. Under this management plan, pesticide use is reduced by 50 percent to 75 percent and improves the soil's ability to hold water and reduces erosion problems.

Another way to improve soil composition is to use manure and to plow under nitrogen-fixing crops as fertilizer. Crops can be rotated every year to prevent cycles of pests from developing and to keep the soil fertile. IPM is a profitable long-term solution to large-scale pesticide use, but it is time consuming. It takes about three years to reach a balanced system.

Enriching the soil near plant roots through the use of natural compost or mulch provides a vast array of nutrients. Chemical fertilizers mostly contain nitrogen, which makes plants look greener and grow faster but does not improve their resistance to disease and extreme heat or cold.

Another way to reduce our dependence on chemicals is to introduce insects such as wasps, ladybugs and flies that are known natural predators of undesirable bugs. Ladybugs and praying mantises will devour aphids, and some wasps kill crop-eating moths. Weeds also can be biologically controlled by subjecting them to deadly plant bacteria or viruses that do not affect the desired crops.

Using natural predators provides several benefits. Buying predator insects or bacteria is much cheaper than the cost of synthetic pesticides. Also, these biological control techniques do not endanger the environment, wildlife or humans. But relying on natural predators takes time and it is difficult to find large quantities of predator insects and plant disease agents.

Another option is to introduce male insects that have been made sterile in the laboratory with radiation and allow them to mate with females in a field that has become overrun with pests. In time, the technique should kill off the unwanted pest population by inhibiting reproduction. This pest elimination method works best if the insect population has already been reduced by the weather or other factors and is isolated from a new wave of nonsterilized male insects.

Insect sex attractants, called pheromones, can lure pests into poisonous traps or natural predators to an infested crop area. This method is highly effective for adult insects, but it is often their young (such as leaf-eating caterpillars) that cause the most damage to plants. Pheromones are available for fewer than 50 species of insect pests.

Food irradiation is being tested as a method to prevent insects from infesting stored food products after they have been harvested. It also makes perishable foods last longer and kills parasites and bacteria, such as salmonella, that cause food poisoning. During the irradiation process, food is passed under radioactive gamma rays, which destroy insects and bacteria without having any radioactive effect on the food. However, the federal Food and Drug Administration has not approved irradiation for widespread use. Opponents worry about chemical compounds that have been found in irradiated food that do not show up in nonirradiated products. Also, irradiation does not kill the bacteria spores that cause botulism but removes their telltale odor.

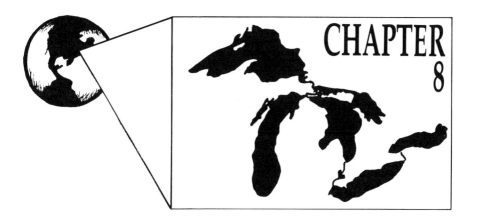

Dangers At Home

While most people are concerned about Michigan's hazardous waste problem, many have not yet made the connection between those frightening barrels of toxins buried in the ground and the chemicals in garden fertilizers, bug sprays, cleaning products and drain openers that most of us use regularly.

We all have contributed to the millions of pounds of hazardous waste that are generated in the Great Lakes basin every year by creating a demand for them as consumers. Before we buy something, we should consider the effluent produced during manufacture, the product's intended use, its disposal and decay. The old-fashioned swat and slap method may begin to look more appealing than a can of chemicals.

The average American household stores three to 10 gallons of hazardous chemicals in the garden shed, garage, kitchen or basement. The cumulative effect is staggering. For example, if each household in a city of 160,000 people (the size of Flint) threw away a small amount of hazardous waste, the city would discharge an estimated six tons of toilet bowl cleaner, 22 tons of liquid household cleaner and five-and-one-half tons of motor oil each month.

Many household chemical products meet the legal definition of hazardous waste—waste that injures or threatens lives—but are treated as harmless municipal solid waste and are buried in community landfills that are not equipped to prevent leaks.

Pouring chemicals down the drain will carry them to a septic tank, where they will drain out under the soil and potentially contaminate nearby waters or drinking wells. In cities, chemicals that are poured down the drain or storm sewers will often pass through sewage treatment plants unchanged, to be discharged directly into a lake or river because most municipal water treatment plants only remove biological wastes from sewage. Once in a lake, the chemicals can enter a nearby water intake pipe and contaminate a community's drinking water and endanger water-dwelling wildlife.

There are simple solutions—take chemicals to a nearby household hazardous waste collection site, recycle used motor oil and batteries or quit using them.

We also use too much pesticide and synthetic fertilizer on lawns and gardens. Lawn keepers and golf courses use more pesticides per acre than do farmers, although farm use accounts for 75 percent of all pesticide use in the United States. Yard chemicals threaten birds by contaminating the worms they eat and birds have mistaken fertilizer granules for food. Eating only five diazinon particles is enough to kill a blackbird or a sparrow.

A 1990 International Joint Commission report on the Great Lakes concluded that "there is a threat to the health of our children emanating from our exposure to persistent toxic substances, even at very low ambient levels...The dangers posed to the ecosystem, including humans, by the continuing use and release of persistent toxic contaminants are severe."

Long-term exposure to the persistent hazardous chemicals that pervade our state's lakes, rivers, soil, drinking water, food and air can inevitably lead to very serious health problems. Substances such as PCBs, dioxins and furans, DDT, lead and mercury can cause liver and kidney failure, central nervous system disorders, cancer, genetic birth defects and miscarriages.

Irresponsible use and disposal of hazardous substances has made the Great Lakes basin a risky place to live. Those of us who cherish the positive things that our home has to offer must fight to eliminate the ominous threat that hazardous chemicals inflict on our lives and the lives of those who will follow us.

As consumers, we can influence the amount of hazardous materials that are produced by not buying any. We can reduce the health risks of the chemicals we do use by properly disposing them.

SOLUTIONS

Things to do:

• Look for substitutes and cut back on products that contain hazardous chemicals. Read labels and follow directions carefully.

• Use latex rather than oil-based paint. Donate unused paint to a school or theater troupe for making sets.

• Don't rinse paint brushes, rollers and pans outdoors—it can pollute the groundwater.

• Use pesticides in your yard sparingly and only when absolutely necessary. Use compost and mulch instead of synthetic fertilizers.

• Try synthetic sex hormones instead of chemicals and poisons to trap indoor pests such as cockroaches and moths. Use old-fashioned mousetraps to catch rodents.

• Don't mix household hazardous wastes, they can be flammable or explosive in the wrong combinations.

- Never pour contents of an original container into an unmarked container.

- Don't pour antifreeze on the ground, it's very toxic.

- Take household and automotive chemicals to a hazardous waste collection center. This is essential if you live in an area that burns its waste in an incinerator.

- Try to avoid bleached white paper products. They create hazardous waste when they are manufactured and when they are disposed of in an incinerator.

- Avoid prepackaged microwave products such as microwave popcorn and pizza, which contain a device called a heat susceptor. Studies by the Food and Drug Administration show that they cause toxic chemicals to migrate into the food they heat.

- Use citrus sprays or put brewer's yeast and garlic in your pet's food instead of using flea collars. Some brands expose your pet to toxic chemicals that can cause liver problems, cancer and birth defects.

- Instead of ammonia-based cleaners, mix vinegar and salt for shiny surfaces, and baking soda, salt and water for tub and tiles.

- To replace spray or solid air fresheners, try potpourri or scented candles.

Things to do for your community:

- If your community does not have a household hazardous waste collection program, help start one.

- If your community has an incinerator, make sure batteries, paints and other materials are removed first.

- Make sure that pollution enforcement agencies in your area are adequately funded and staffed.

- Keep an eye on local industries. If they are polluting, pressure them to stop and clean up their messes.

Contacts:

- Michigan Department of Natural Resources, Waste Management Division, Box 30038, Lansing, Mich. 48909. It provides brochures and a list of hazardous waste collection days and centers across the state. It can help set up hazardous waste collection programs. Call: (517) 373-2730 or the household hazardous waste agent in your area. Within (313), call 1-800-468-9612; within (517), call 1-800-468-8879; within (616), call 1-800-468-8896; and within (906), call 1-800-624-4100.

- To find out where to recycle used motor oil in your area, call 1-800-634-9504.

- The National Wildlife Federation's Great Lakes Natural Resource Center publishes the pamphlet, "Do You Have a Zero Discharge Home?" which contains information about how to eliminate hazardous chemicals from the home and provides a chart about non-toxic alternatives to typical household chemicals. 802 Monroe, Ann Arbor, Mich. 48104, (313) 769-3351.

- Send for a free copy of the "Citizen's Guide to Toxic Substances," National Wildlife Federation, 1412 Sixteenth Street NW, Washington, D.C. 20036.

- The International Joint Commission's booklet, "Hazardous Wastes from Homes," describes the history of wastes, different hazardous waste treatment facilities, health risks and the disposal of several common household chemicals. Write: IJC, Great Lakes Regional Office, 100 Ouellette Ave., 8th Floor, Windsor, Ontario, N9A 6T3.

- Call the DNR Hazardous Waste Hot Line to report improperly managed hazardous waste disposal sites, 1-800-621-3191.

- Call the DNR Pollution Emergency Alerting System (PEAS), to report environmental law violations (illegal dumping, etc.) 24 hours a day, 1-800-292-4706 or the EPA Pesticide Hot Line: 1-800-858-7378.

- The EPA's Emergency Planning and Community Right-to-Know Information Hot Line will help you find out which companies pollute in your area. Call 1-800-535-0202.

- Call the EPA's Resource Conservation Recovery Act/Superfund Hot Line: 1-800-424-9346.

Further reading:

The Non-Toxic Home, by Debra Lynn Dadd, $11.95. Available in bookstores or write: Box 1506, Mill Valley, Calif. 94942.

The Chemical Free Lawn, by Warren Schultz, Rodale Press (1989). Contains information about which types of grass are best to plant in different regions, fertilizing, watering and natural pest-control techniques.

The Encyclopedia of Natural Insect and Disease Control, Rodale Press, 33 East Minor Street, Emmaus, Penn. 18098. Also in bookstores.

For information about alar, read "Bad Apples" by Consumer Reports, May 1989.

Intolerable Risk: Pesticides in Our Children's Food, $5, from the Natural Resources Defense Council, 740 West 20th St., New York, N.Y. 10011.

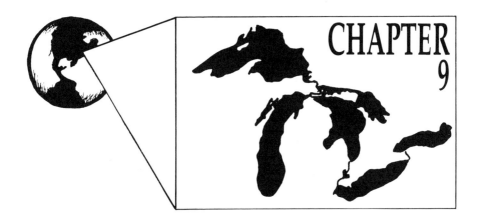

Invisible Dangers

ne of the most basic types of air pollution, called suspended particulate matter, consists of particles of dust, soot, lead and cadmium. All parts of the state meet federal standards for particulate matter except Wayne County, according to DNR.

Citizens who frequently burn wood in their fireplaces and wood stoves or burn leaves in the fall contribute to particulate air pollution. People who burn plastics, tires, paint and motor oil send clouds of poisonous chemicals into the air. Many of these items can be recycled or taken to a local household hazardous waste collection site instead.

More than three-fourths of the 120 million pounds of hazardous pollution released in Michigan during 1988 was sent into the air, according to the EPA. The Upjohn Company in Portage, the state's number one air polluter in 1988, emitted 7.4 million pounds of airborne chemicals. According to several air quality experts, Michigan has one of the weakest state air pollution laws in the nation.

When Michigan's Air Pollution Control Act was passed in 1965, it was quite progressive and beat the federal Clean Air Act by five years. But 25 years later, critics contend that our state's air pollution law is outdated and badly in need of reform. Even state air quality officials acknowledge that the aging law lacks an effective legal mechanism to force polluters to control their emissions and does not contain any formal regulations for airborne toxic pollutants. On the plus side, since 1972 the Michigan Air Pollution Act has reduced by more than 60 percent power plant sulfur dioxide emissions, which contribute to acid rain.

Most of the hazardous substances that taint the Great Lakes and many of Michigan's inland lakes originated as air pollution. When contaminants such as lead, mercury, PCBs and dioxins are sent into the sky by manufacturing plants and incinerators, they mix with water vapor in the clouds. Precipitation carries the pollutants into lakes and streams and spreads them over the soil. This process is called atmospheric deposition. Research has shown that 97 percent of the lead and 90 percent of the PCBs that enter Lake Superior were carried there by the wind. Much of this airborne pollution originated many miles away from the Great Lakes basin.

Since World War II, more than 60,000 different chemicals have been developed for use in business and manufacturing. Since 1970, the EPA has only regulated seven toxic air pollutants under the federal Clean Air Act.

In Michigan, though, the DNR began in 1980 to control new or modified sources of air pollution by withholding permits until facilities proved that their emissions would not harm human health or the environment. New sources must install the best pollution control technology available and are subject to periodic DNR inspections to determine if more emission control is necessary. Michigan is the first state in the nation to develop such a comprehensive program for dealing with air toxins.

But DNR air pollution enforcement officials still lack a system to force polluters to stop discharging hazardous wastes unless the department can prove that they are posing an extreme threat to human health. Under the present system, once a permit is issued, it remains until someone brings a complaint and proves damage. There is no revision or review. Permits that were granted 20 years ago are still legitimate, despite their obsolescence.

The biggest, and most far-reaching air pollution problem in Michigan is acid rain. Michigan has the third highest level of rainfall acidity in the country. Acid rain is created when sulfur dioxide and nitrogen oxides—produced by coal-burning utilities and automobiles—mix with moisture in the clouds and fall to the ground as sulfuric and nitric acids via rain, snow, ice, hail, wind, fog or rain.

The relative acidity of a substance is measured by its level of pH. The scale ranges from 0, the most acidic, to 14, the most basic, or alkaline. Distilled water has a neutral pH of 7. Because natural rainwater has some carbon dioxide dissolved in it, it is slightly acidic with a pH of 5.6. Vinegar has a pH level of 3, and lemon juice is measured at 2.3. Rain in Michigan has an average pH level of 4.2, a level that is considered harmful.

The problem persists in a vicious circle. To eliminate local ground-level pollution, power plants in Ohio, Michigan and Ontario built tall smokestacks that send pollutants high above urban areas into the atmosphere. But as Isaac Newton said, what goes up, comes down. Satellite experiments have shown that sulfur dioxide particles originating from the Ohio River Valley are borne northeast by prevailing winds into the Adirondack region of New York and parts of Ontario and Nova Scotia, where they descend on the land as acid precipitation.

Some scientists believe some of the contaminants found in the Great Lakes, such as the pesticide DDT (dichloro diphenyl trichloroethane), may have come from as far away as Central or South America. In 1980, an insecticide called toxiphene, which is used on southern cotton fields, was found in an isolated part of Lake Superior.

Much of the soil in Michigan's Upper Peninsula contains acidic sand or granite and cannot neutralize acid precipitation as well as places that have limestone or other alkaline elements. Lakes and streams lined with acidic soil also lack the ability to neutralize acid rain. As a result, more than one-third of the lakes in the Upper Peninsula are vulnerable to the ravages of acid rain. McNeary Lake in the Hiawatha National Forest has become so acidic, all of the fish in it died.

As freshwater lakes, which normally have a pH of about 8, become more acidic, smaller aquatic plants and organisms start to die. Organisms that relied on them for food, such as minnows and crayfish, also die, making food sources scarce for large fish, frogs and other amphibians. The problem worsens in the spring because of the influx of melted acid snow or water runoff. This process, called acid shock, introduces into the lake melted snow that is 10 times as acidic as normal water at the worst time of the year—when many species of fish, frogs and toads are spawning.

Acid water has been shown to cause skeletal deformities or death in trout and salmon embryos. It also causes toxic metals such as lead, mercury and aluminum, which clogs fish gills, to leach out of bottom sediments and nearby soil. At a pH level of 4.5 or less, fish, frogs and insects will be killed, impacting the rest of the food chain.

Acid rain also kills trees and other vegetation. Acidic rainfall and accumulations of ground-level ozone break down the waxy coating of leaves and pine needles, weakening a tree's ability to prevent water loss and damage from pests, extreme cold and drought. Losing the protective coating inhibits the tree's growth and nutrient uptake, causing its leaves and needles to turn brown and fall off. Eventually, disease and toxic metals leached from the soil attack the roots and the tree dies. Long-living pines are especially susceptible to acid rain damage because their needles are constantly exposed to pollution.

Widespread tree loss in Europe is blamed on acid precipitation, especially in West Germany, where an estimated 35 percent of forest land is already dead. In the United States, large stands of pines are dying at high elevations of the Rocky Mountains and the Adirondacks. Studies indicate that the decay will soon spread to lower areas.

Our forests are invaluable resources. They protect the soil from erosion, support a huge array of wildlife and provide an aesthetic refuge from urban chaos. When trees begin to fall from acid rain damage, so will our lumber, construction and paper industries.

Agricultural crops are being damaged by acid rain, too. Some crop yields are reduced and other crops, such as apples, become so splotched and damaged by the acid that they cannot be sold. Researchers also believe that acid rain kills beneficial bacteria and microorganisms that break down wastes and keep the soil moist and rich.

Acid rain also is destroying society's greatest monuments. The Statue of Liberty, the Washington Monument and Independence Hall in Philadelphia are showing signs of acid corrosion. The Parthenon, Taj Mahal and ancient Mayan temples in Mexico are rapidly aging and noses and ears are falling from classic marble busts in Italy. The royal palace in Amsterdam and venerated medieval buildings in Poland also are endangered. On a smaller scale, acid rain corrodes bridges and eats car paint.

Michigan already has cut sulfur dioxide emissions from power plants by 60 percent since 1972 by requiring the use of clean-burning low-sulfur coal and smokestack scrubbers that neutralize the gas before it escapes. However, nitrogen oxide emissions have decreased only 10 percent since 1972. Because most nitrogen oxide comes from car exhaust, regulation of the millions of car owners is nearly impossible.

Fishless lakes and stark, dead forests could destroy the beauty and tranquility of Michigan and cripple the Great Lakes region's multibillion dollar fishing and tourism industries. To avoid this environmental and economic devastation, Michigan has to stop sulfur dioxide and nitrogen oxides emissions. But because acid rain is a regional problem, we can only be effective if we work together with neighboring states and provinces. Many of the major sources of pollution that cause acid rain are in our own midwestern backyard. We have a responsibility to stamp it out.

Another issue that has to be addressed globally is the production of ground-level ozone.

People often confuse the two types of ozone. The natural kind of ozone, which consists of three oxygen atoms bonded together, is the substance that makes up the protective ozone layer in the stratosphere. In the lower atmosphere another type of ozone forms when nitrogen oxides and hydrocarbons, produced by cars and industries, react in the presence of warm sunlight. Other sources of hydrocarbons, also called volatile organic compounds, include paint, solvents and dry cleaning chemicals.

Ground-level ozone, which exists as a gas, is a major component of urban smog. Like acid rain and air toxins, ozone is an air pollution problem that does not obey political boundaries. For example, one of the state's biggest problem areas for ozone is in southwestern Michigan, where ground-level ozone floats across Lake Michigan from Milwaukee and Chicago. As a result, the region frequently exceeds federal air quality standards for ozone, even though most of the pollution was created in another state. The southeastern part of the state, particularly Detroit and its suburbs, also has consistently high levels of ozone. Eleven counties in these two areas of Michigan exceed federal standards for ozone, according to the EPA.

Reducing urban ozone and other types of air pollution will require a societal shift toward cheap, convenient public transportation. As the country's population continues to increase, more and more cars will be added to our already overcrowded freeways. More cars will be idling in traffic jams, sending clouds of exhaust fumes into the air, perpetuating ozone, acid rain and global warming problems. Using cleaner-burning fuels and designing more fuel efficient and pollution-free cars are some other alternatives that need to be explored. Because of the auto industry's powerful presence in Michigan, a shift away from the daily use of cars by Michigan's five million citizens is unlikely.

The short-term effects of ground-level ozone are familiar to anyone who has walked through a big city on a hot, sunny day with very little wind. The air irritates mucous membranes, causing people to cough, choke and have difficulty breathing. Elderly people with heart and lung problems, infants and people who breathe through their mouth, bypassing the nose's protective filtering mechanism, are more vulnerable to the harmful effects of ground-level ozone.

Constant, long-term exposure to ozone and other air pollution reduces the body's ability to remove pollutants from the lungs. Microscopic particles can penetrate the alveoli, or tiny air sacks in the lungs, and increase the chance of lung cancer or respiratory infections. It also can reduce resistance to colds and exacerbate emphysema, heart disease and bronchitis. Americans spend $40 billion to $90 billion for health care costs associated with urban smog, according to the American Lung Association.

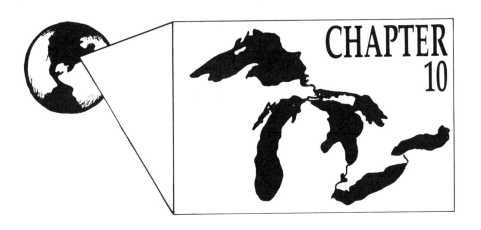

Indoor Air

The air indoors can be worse than ozone-laden urban air. Studies by the EPA have found that the air inside a home or office building can be up to 100 times more polluted than outdoor air. Because people spend so much of their lives inside their homes and offices—many of which are poorly ventilated and emit toxic fumes from various materials—some scientists believe that indoor pollution may be one of the top 10 causes of death in the United States.

Smoking is the country's number one cause of cancer. "Passive smoke" that is inhaled by nonsmokers may increase their chances of getting lung cancer by as much as 30 percent, according to the U.S. Surgeon General. Studies by the EPA indicate that benzene levels are 30 percent to 50 percent higher in the air of homes where people smoke than in nonsmoking households. Benzene, a known human carcinogen, is one of the 4,600 known chemical components of cigarette smoke.

Materials used to construct homes and other buildings sometimes cause indoor pollution as well. Some types of inexpensive pressed-wood lumber emit formaldehyde, which can cause headaches, nausea and irritation of the eyes and respiratory tract. It is a probable cancer-causing substance in high concentrations. Formaldehyde also is found in plywood paneling, particle board, flooring materials, cabinets and furniture. Solid wood products are a better alternative, according to the EPA.

Many people in Michigan have had their homes "super insulated" to minimize heat loss during our cold winters. Although this process has saved home owners money, it may cost them their health. Those triple-sealed windows and airtight weather-stripped doorways trap harmful pollutants from gas appliances, fireplaces and cigarettes.

Other products that contribute to indoor pollution include paints, solvents, pesticides, chemical cleaners, home permanents, nail polish and nail polish remover. Furniture fabric, carpets, drapes, air conditioners and humidifiers can cultivate microscopic bacteria, dust, mold and fungi. Draperies and dry-cleaned clothing also emit formaldehyde. Kerosene heaters and poorly ventilated gas stoves give off nitrogen dioxide, carbon monoxide and fine particulates, which can cause respiratory problems.

Radon is another source of air pollution inside schools, offices and homes. The colorless, odorless gas is created when natural deposits of uranium in rocks and soil decompose and release radon gas. In most places, radon gas is present in harmless amounts.

Radon only becomes a threat when buildings are built over uranium deposits, trapping the gas and exposing the people who live and work there to unhealthy concentrations of radon.

The EPA estimates that one out of every eight American homes is contaminated by radon gas. Radon occurs in spotty patches almost everywhere in Michigan, but there are a few particularly hot spots. Several counties in the Upper Peninsula, including Marquette, Iron and Ontonagon, have high levels of radon because the area has a thin layer of soil over a slab of granite, which tends to contain uranium. Although the lower part of the state has several deep layers of different soil types covering the uranium-holding granite, houses built in Washtenaw, Wayne and Jackson counties have a 50 percent chance of radon contamination. Geologists think radon may migrate from uranium deposits to the upper layers of soil through air pockets.

Because radon gas seeps through cracks in foundation walls, basements and ground-level rooms usually have the highest levels of contamination. Radon levels rise in the winter because the buildings are sealed against the cold.

When radon gas decays, it splits into tiny particles called radioisotopes, which adhere to dust particles and can be inhaled. Radon radioisotopes are tiny enough to enter individual lung cells, exposing them to unusually high amounts of radiation, which can cause lung cancer if the exposure is over a long period. Radon is a leading cause of lung cancer in the United States, second only to cigarette smoking.

Radon also can contaminate nearby groundwater and drinking wells. Studies have shown that it may actually be more harmful to shower in radon-contaminated hot water than to drink it because as water molecules evaporate into the air they carry radon particles with them, which can be inhaled. Similarly, cigarette smoking in a radon-contaminated building may increase the risk of cancer associated with long-term exposure to radon gas.

The EPA has set the safety limit of radon at four picocuries (a unit of radiation) per liter of air. If the level exceeds that, corrective measures should be taken. This could be as simple as adding more windows or creating better ventilation with fans.

But for homes with higher levels of radon, fixing the problem can be somewhat expensive. One method is to install a network of pipes under the foundation to draw most radon away from the building and vent it outside where it dissipates safely. Depending on the type of foundation slab, this costs between $1,500 and $4,000. Special carbon filters can be added to eliminate radon from well water.

Another indoor health hazard is asbestos, a fire-resistant mineral that has been widely used since 1900. Tons of asbestos were sprayed onto the ceilings and walls of schools for fireproofing and soundproofing. Asbestos fibers have been wrapped around pipes and heating ducts, mixed into insulation, roofing and flooring materials and woven into flame-retardant curtains, blankets and protective clothing for fire fighters. It is still used in automobile brake and clutch linings.

As asbestos ages it crumbles, sending a lethal cloud of tiny asbestos fibers into the air. If inhaled, even trace amounts of these fibers can cause lung cancer. More than 30 million tons of asbestos were used between 1900 and 1986 and the EPA estimates that asbestos is responsible for 3,000 to 12,000 cases of lung cancer every year in the United States.

Cleaning up asbestos is dangerous, expensive and difficult. In 1984, Congress provided $600 million to struggling school systems under the Asbestos School Hazard Abatement Act, but only a small portion of this money has been appropriated. EPA regulations under the Toxic Substance Act only require that schools be inspected for asbestos contamination. If hazardous amounts are discovered, the EPA requires the school to notify the community, but it does not have the power to force the school system to undertake a costly asbestos reduction project.

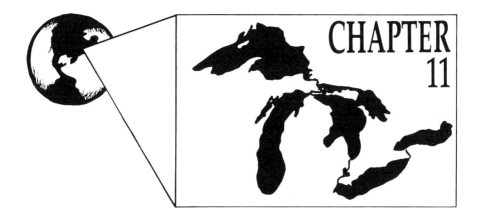

How Does Michigan Penalize Polluters?

Michigan's Air Pollution Control Act was written before the federal Clean Air Act and is not as strict as federal air regulations or other states. Its provisions are either skeletal or nonexistent, as in the case of toxic air emissions. Under the current law, there is no way to stop existing facilities from polluting the air with toxic chemicals without proving damage and there are no formal rules that prevent new sources of pollution.

The Michigan law's cumbersome enforcement mechanisms seem to tie officials' hands instead of empowering them to bring polluters to justice. The DNR can only enforce regulations with criminal penalties if parties violate their permits or disobey an administrative order to install better control technology. In a criminal proceeding, the DNR has to prove beyond a reasonable doubt that a company was willfully violating the law.

Penalizing a polluter through civil rather than criminal proceedings takes less time, more frequently leads to success and enables the DNR to pursue more cases. Civil prosecution, which just requires the proof to tip one way, enables the agency to pursue "equitable relief," or force a company to pay for the damage caused by its emissions. This is not possible under the criminal system, which only punishes polluters through fines, imprisonment or closure.

The updated version of the federal Clean Air Act gives states the ability to enforce air pollution regulations both civilly and criminally. This allows for greater flexibility and a wider range of penalties. For example, the state could force a company to pay for cleaning up the contamination it caused through civil proceedings and force its officers to pay a fine or spend time in jail.

Why has the Michigan Air Pollution Act remained so weak for so many years? One answer is strong representation from the state's powerful automobile and chemical industries, which have a vested interest in keeping the law lenient and are willing to pay for a strong corps of lobbyists in Lansing. The air pollution law is a powerful factor in determining whether industries will find it profitable to do business in Michigan. Therefore, industry has strongly opposed any attempts by the government to strengthen the act. On the other side, there has been little citizen involvement in air quality issues. But now that the federal act has improved, state lawmakers are expected to update state laws.

Meanwhile, the Great Lakes remain a hotbed of hazardous waste. The region produces or consumes approximately 30,000 different chemicals and generates 32 percent of U.S. hazardous wastes and 41 percent of Canadian wastes. Michigan industries alone dumped 120 million pounds of hazardous chemicals into the air and water in 1988, placing the state 10th among the 50 states in toxic pollution, according to an EPA study. Other Great Lakes states, such as Ohio and Indiana, had even more pollution: 229 million and 212 million pounds, respectively. In Michigan, we have more than 2,800 known contaminated sites in need of cleanup, 79 federal Superfund sites and countless other abandoned toxic dumps that have yet to be discovered.

So far, fewer than 30 of the sites on the state's catalog of contaminated spots have been deemed clean enough to be scratched from the list. DNR officials say it will take $3 billion and many, many years to detoxify Michigan's environment because for years the department's staff was solely underfunded. It could not keep a thorough inventory of contaminated sites, let alone clean them up. A shrinking budget slowed them down again.

Act 307, passed in 1982, requires the DNR to list contaminated sites. Often times, the bad publicity of being placed on the list is enough to coax companies to voluntarily clean up the mess. If that doesn't work, the state can demand cleanup through recent legislation that holds polluters responsible.

As funding has increased over the past 11 years, the DNR has been able to expand the number of hazardous waste sites they have identified from 268 to nearly 3,000. The more they are able to search, the more they find. With a 1988 bond approval, chances are good that more progress will be made to erase some of the thousands of contaminated sites from the Act 307 list.

The 79 Michigan sites on the federal Superfund list are eligible for a portion of the program's $8.5 billion budget. Michigan has the fifth highest number of Superfund sites in the country behind New Jersey, Pennsylvania, California and New York.

Southeastern Michigan, the most populated part of the state, suffers the most hazardous contamination. Sources of contamination include leaking landfills and dumps, faulty underground gasoline, oil and propane storage tanks, dry cleaning chemical leaks and hazardous spills from trucks, trains and ships. Major Michigan industries, such as steel companies, automobile manufacturers, and chemical and plastics companies, unload the most hazardous chemicals into the environment. The chemical industry alone accounts for more than 70 percent of all of the hazardous waste in the United States.

Industries that produce, treat and store hazardous materials are regulated by Michigan's Hazardous Waste Management Act, passed in 1979. It requires a company to apply for a permit to generate, treat, store or dispose of hazardous wastes. Currently, there is only one commercial hazardous waste landfill in Michigan that is licensed to accept wastes created by other companies, the Wayne Disposal in Belleville. Industries that do not take their waste there must store it themselves or ship it out of the state.

Before 1979, many industries just poured their wastes into a pit in the ground called a lagoon or surface impoundment. The hazardous chemicals quickly seeped into the ground and threatened drinking water supplies. Now regulations require that surface impoundments must be lined with a thick layer of plastic and fitted with leak collection and monitoring systems.

Hazardous waste landfills are subject to increasingly strict design standards. Any hazardous substance that may leach into the ground is prohibited from hazardous waste landfills without first undergoing neutralization and solidification. The landfills must be built over a layer of impermeable clay soil and double or triple layers of plastic liners that are about 30 times as thick as a plastic garbage bag. A system of pipes collects any leachate and carries it to a hazardous waste treatment facility. Landfill operators must continuously monitor the structure for leaks and prove that they can pay for cleanup if contamination occurs.

Industrial waste incinerators must be fitted with sophisticated pollution control devices. In theory, incineration detoxifies complex chemical compounds by breaking down their molecules under intense heat and transforming them into harmless gases. Although this disposal method is effective for most kinds of hazardous waste, it is extremely expensive and represents a formidable health hazard if proper controls are not in place.

But one industry's waste can be another company's raw materials. Hazardous waste reuse or recycling is an economical and environmentally sound disposal alternative that very few companies in Michigan utilize.

In some parts of Europe, waste clearing houses have been set up to coordinate the exchange of chemicals between industries that want to get rid of them and those that need a particular substance for their manufacturing process. European firms have reduced the hazardous waste stream by as much as one-third. The idea is catching on in the United States and several exchanges have cropped up. For example, 3M, a Minnesota company, sells ammonium sulfate, produced as a by-product during videotape manufacturing, to fertilizer companies.

The EPA estimates that 20 percent of the hazardous chemicals currently produced in the United States could be recycled, far more than the tiny percentage now reused. Reusing and recycling chemicals saves money in raw materials, cuts down on the amount of harmful by-products that are created during manufacturing and eliminates the threat of landfill or incinerator pollution.

As the cost of hazardous waste disposal continues to climb, more waste producers may be inclined to "economize" by dumping their wastes illegally. Some common methods include dumping the hazardous waste into rivers or sewers, dumping it or burying it on an isolated piece of land or letting it spill out onto the road from a moving truck.

In one instance, 20 drums of a hazardous solvent were dumped into a ditch along the road in Allegan County, south of Grand Rapids. The drums were traced to a Grand Rapids company, which allegedly paid someone $100 to take the drums, according to investigators in the DNR's Law Enforcement Division. Hiring a licensed waste hauler to transport the drums would have cost more than $3,000. The Grand Rapids company and its president and vice president have been charged with 10 violations of the Hazardous Waste Management Act. Each charge carries the penalty of up to a $25,000 fine and a year in jail.

Enforcement officials say many of their tips about illegal dumping come from employees of hazardous waste-generating companies who witness an illegal act or are asked to commit a crime. Usually, they reach the law enforcement division through the DNR's 24-hour, toll-free Pollution Emergency Alerting System (PEAS) or the Hazardous Waste Hot Line. To encourage more employees to come forward, DNR officials say they are extremely careful to protect the informer's identity because his job may be on the line.

Besides responding to calls from citizens, hazardous waste regulators also try to eliminate illegal activities by randomly inspecting hazardous waste transportation vehicles and check for proper documentation. Hazardous waste permit holders must account for their waste "from cradle to grave" through a series of manifests, or records, on how much each generator produces and track every move a hazardous substance makes during its disposal.

By looking through these manifests, DNR personnel can pinpoint potential law breakers. For instance, if they know that a particular industry produces a certain amount of waste each month, but have not received any manifests for the disposal of that waste, they can ask law enforcement officers to investigate.

While this system is effective, its impact is diminished by a lack of state funds. Only three staff members keep track of manifests. The Law Enforcement Division has only 11 officers who are responsible for tracking down pollution violations all over the state, not only for hazardous waste dumping, but also for oil and gas contamination, improperly operated landfills and waste transportation vehicles, and water pollution. In 18 months, DNR investigators brought 82 cases before the state attorney general. Properly staffed enforcement teams is a critical need in the state, even as the state's budget is cut.

SOLUTIONS
Things to do:

• Drive less. Walk, bike, car pool or take public transportation instead.

• Regularly check your car's pollution control equipment.

• Make your home more energy efficient. Caulk windows, add insulation and storm windows and doors.

• Learn about radon contamination in your area by calling the county health department. If your county has large amounts of uranium deposits, test your home for contamination.

• If you are buying property, have it tested for radon before you move in or start building. An initial screening costs between $12 and $40.

• If you live in an older home, check for asbestos around pipes and other areas. If you find it, hire a professional asbestos abatement company to safely remove it. Do not try to do it yourself. Asbestos fibers are extremely dangerous even in small amounts.

• If you are building a new home or adding on to your house, find a contractor who uses environmentally safe building materials.

• Minimize use of pressed wood products, which contain formaldehyde. Solid wood products are a better alternative.

• Quit smoking. If you are not a smoker, ask others to smoke outside to avoid indoor air pollution.

• If you have a gas stove or kerosene heater, make sure the room where they are used is well-ventilated.

Things to do for the community:

• Make sure local governments issue health warnings when ozone levels in your area exceed federal standards.

• Attend the Michigan Air Pollution Control Commission's monthly meetings and voice your concerns about clean air and Michigan air pollution regulation.

• Join a group, such as the American Lung Association or Michigan United Conservation Clubs, which is working to strengthen Michigan's air pollution legislation.

• Urge the schools in your community to test for radon and airborne asbestos.

• Urge your federal legislators to support strong acid rain and clean air legislation.

Contacts:

• For information about Michigan air quality, contact the Michigan Department of Natural Resources Air Quality Division, Box 30028, Lansing, Mich. 48909, (517) 373-7023, or call your county health department.

• Contact the American Lung Association of Michigan, 403 Seymour, Lansing, Mich., (517) 484-4541, for information about Michigan air pollution, legislation and health effects.

• For information about combating acid rain, contact the Izaak Walton League, 1701 North Fort Meyer Drive, Suite 1100, Arlington, Va. 22209, (703) 528-1818.

• Write the Acid Rain Information Clearinghouse at 33 South Washington St., Rochester, N.Y. 14608, (716) 546-3796.

• For information about getting a health hazard evaluation of your workplace, call the National Institute for Occupational Health and Safety, 1-800-35NIOSH.

• To find out about federal funds to help schools pay for asbestos abatement, call the Asbestos School Hazard Abatement Act hot line, 1-800-835-6700.

• Toll-free Cancer Information Service 1-800-426-4237.

Further reading:

• To order a copy of "Acid Rain: What it is—How You Can Help," write National Wildlife Federation, 1412 Sixteenth St. NW, Washington, D.C. 20036.

• The EPA offers three free booklets with information on indoor air pollution, ''The Inside Story: A Guide to Indoor Air Quality,'' ''A Citizen's Guide to Radon,'' and ''Radon Reduction Methods: A Home Owner's Guide.'' Write U.S. Environmental Protection Agency, Office of Public Affairs, Washington, D.C. 20460.

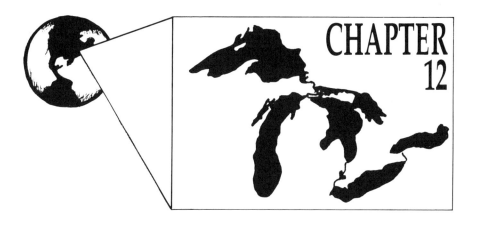

Habitat Destruction

Humans' assault on wildlife is most apparent in their consumption of animal habitat.

If a time machine could take you back to the 17th century when Europeans first discovered Michigan and the Great Lakes, our industrialized state would look dramatically different. Instead of cottages and summer homes, lake shoreline would be covered with coastal wetlands filled with water-loving plants such as willowy bull rushes and cattails, colorful orchids and wildflowers and carnivorous pitcher plants and sundews. Lakes would be free from the drone of speeding power boats. You might see some caribou, wolverines and trumpeter swans, wildlife that have been extirpated in Michigan. Animals still common to Michigan would be around in greater numbers.

The French explorer Samuel de Champlain, one of the first Europeans to set foot in Michigan and the Great Lakes basin, was dazzled by the copious array of flora and fauna that existed among the rich lands.

"This country is so very fine and fertile that it is a pleasure to travel about in it," he said in his correspondences. In his diary, he reported seeing "a great quantity" of swans, geese, cranes, ducks, teal, larks and "other kinds of fowl too numerous to count." Champlain found Lake Huron, which he named the Freshwater Sea, to be teeming with "many kinds of excellent fish...principally trout, which are of enormous size; I have seen some that were as much as four and a half feet long and the smallest ones are two and a half feet in length. Also pike of like size and a certain kind of sturgeon, a very large fish, and marvelously good to eat."

Champlain sketched the various fruits and nuts he encountered in the Great Lakes region and wrote, "there is an abundance of vines and plums, which are very good, raspberries, strawberries, small wild apples...there are also quantities of small cherries and wild ones (black cherries)." He recorded that he saw many oaks, elms and beech trees. He saw rabbits and partridges and hunted the plentiful population of deer with his Indian guides.

Although Champlain clearly appreciated the natural riches of the land he explored, he believed it was there to exploit rather than protect. Unlike Indians who saw themselves as part of nature, European settlers taught us to view the land as something to be conquered and transformed. In pre-settlement times, Michigan had as much as 11 million acres of wetlands. To the white man, dark swamps and marshes, with their disease-carrying mosquitoes and humidity, were nothing but a nuisance. They drained them for agriculture and filled them in to build homes and businesses.

That trend continues into the 21st century as we develop more office and housing complexes and expand agricultural acreage to increase food production at the expense of the wildlife that depends on wetland habitats.

Currently, only three million acres of Michigan's original 11 million acres of wetlands are left. For years, people thought economy-boosting industries and homes for people should take precedence over homes for animals and plants. Gradually, people are becoming more aware of the important benefits of wetlands and are more willing to protect the relatively few acres that we have left.

Michigan's wetlands were formed over billions of years during the ice age when the Great Lakes basin was covered with glaciers. As the climate warmed, the icecaps started to melt and retreated to the north. This process occurred four times. With each recession and advance, the glaciers etched a series of holes and craters in the underlying rock. Places with soft, sandy soil were most affected by the glacial activity and thus have the greatest amount of lakes, ponds and wetlands, created when the holes filled with water.

Wetlands come in many different forms. Marshes, bogs, swamps and ponds are all considered wetlands, as is any area where water meets land. This can include a meadow, which may not look particularly wet, but has a high water table and floods during rainy seasons. It also can include a small pool of water that dries up in the summer sun. In general, a wetland is a place where water covers the soil for all or most of the year, or long enough to allow for the development of an ecosystem with plants and animals that need characteristics of both water and land environments to survive.

Because of this dry land/aquatic flexibility, a wetlands habitat supports the greatest diversity of plant and animal species. Consequently, human destruction of wetlands threatens the extinction of many species of plants and animals.

The most common types of wetland in Michigan are marshes, swamps and bogs. Marshes are usually found in shallow, stagnant water beside ponds or lake bays. The water in a marsh is shallow, but deep enough to attract ducks, geese and herons who use the marsh for nesting and raising their young. Grassy vegetation such as cattails, reeds and arrowhead also grow there. In deeper marshes, floating plants such as water lilies, lotus bloom and mats of elodea and pondweed provide an important source of food for young ducklings and other water birds. Fish use shallow areas of marshes for breeding and protection. The habitat also suits frogs, snakes and turtles.

Swamps resemble flooded forests. They contain several types of trees and shrubs with submerged root systems and include willows, cottonwoods, cedar and red and silver maple. Wildflowers, such as the yellow ladyslipper, are typically found in swamps. Many of the animals found in marshes live in swamp areas, as well as tree frogs, salamanders, raccoons and water shrews.

Bogs are formed when vegetation gradually fills in the edges of a glacial lake. Decomposing sphagnum mosses and sedges create an acidic mat of peat moss, which supports unusual plants such as the carnivorous pitcher plant and sundews. These plants trap insects and digest them for the nourishment they are unable to receive from the bog's nutrient-poor soil. However, bogs often are destroyed by peat mining, which is a common practice in the southern parts of the state where there is a high demand for garden potting mixes.

Another type of bog, known as a fen, has a more alkaline soil content due to underlying calcareous deposits. Because fens have more nutrients, they can support many more plants than bogs. One threatened plant, Houghton's goldenrod, only grows in fens in northern Michigan along the shoreline of lakes Huron and Michigan.

Wetlands have important ties to Michigan's sport and commercial fisheries. As the spawning ground, nursery, feeding place and safety zone for most species of fish, shallow wetlands are critical to the success of this important Michigan industry. Similarly, without these watery breeding places, ducks, herons, geese, salamanders, frogs, turtles and many other species of wildlife wouldn't be able to reproduce and would disappear.

Wetlands are vital for preserving plant species diversity. Many plants will only grow in a wetland habitat. Wetlands provide natural, hands-on biological laboratories for university researchers and scholars and classrooms for children learning about their environment. They provide endless hours of recreation and appeal to people who enjoy canoeing, boating, bird-watching, photography, fishing or hunting.

Until recently, many people were unaware of the many ecological services wetlands provide as a vital part of the Great Lakes ecosystem. Coastal wetlands along the Great Lakes buffer shores from strong, gusty waves and prevent property damage. Wetlands absorb and store excess rainwater to prevent flooding and wetland plants catch sediments and pollutants, preventing them from reaching other ecosystems. As part of the hydrologic cycle, wetlands recharge groundwater supplies with surface water during wet seasons, and replenish surface waters with groundwater during dry times. Some communities have taken advantage of these natural benefits by incorporating wetlands into housing developments to prevent flooding and using them as a final filtering stage in their water treatment process.

Since 1976, a portion of Michigan duck hunting licenses has helped protect state wetlands. The duck stamp program has raised almost $2 million, matched by federal funds, to purchase 4,000 acres of wetland habitat in Michigan. Some environmental groups, like the Nature Conservancy, use private donations to buy and preserve wetlands.

Michigan's 1972 Inland Lakes and Streams Act requires permits from the DNR or the U.S. Army Corps of Engineers for any dredging or filling activities in wetlands that are adjacent to bodies of water. The law requires regulators to consider environmental, social and economic factors before issuing a dredge or fill permit.

Wetlands not connected to rivers, lakes or streams were unprotected until the Michigan Legislature approved the Goemaere-Anderson Wetland Protection Act in 1979, which dictates careful management and preservation of all of Michigan's remaining wetlands. The Land and Water Management Division of the DNR must issue permits for projects that would alter any wetland.

Residents can review any application for wetland alteration and can call for public hearings and further review of the project. Michigan United Conservation Clubs has acted as a clearinghouse of information for people interested in monitoring the wetlands activity in their community. The "Wetlands Watch" sends to anyone interested the notices issued by the DNR and the Army Corps about wetland alternation applications. The Michigan Audubon Society recently has begun a similar service.

The Wetlands Protection Act also authorizes local governments to adopt special zoning ordinances to prevent wetland destruction. About 20 Michigan communities have enacted wetland protection ordinances. The Tip of the Mitt Watershed Council in northern Michigan supplies newly-forming community groups with sample wetlands ordinances, information on the legislative process and deciphers dredge and fill permit applications. It also tells people how to request a wetlands public hearing.

Because of Michigan's exemplary use of federal and state wetlands protection laws, the EPA granted the DNR power to enforce Section 404 of the federal Clean Water Act, which concerns development in Michigan waters and wetlands. No other state has been given this privilege. The action facilitated quicker, more localized decision-making and opened the door to greater citizen involvement in protecting Michigan's wildlife resources. To help state and local officials manage these areas more wisely, the DNR has employed a computer to begin to map out all of Michigan's wetlands.

On another front, some ecologists advocate creating wetlands. Other scientists are skeptical, saying people cannot create the fragile and complex wetland ecosystems which took nature billions of years to develop.

But the wetlands that have been preserved are still in danger. Urban and agricultural development has robbed wildlife of its home by polluting their habitats with deadly chemicals. Wildlife is more sensitive to environmental pollutants than humans, but our regulation system concerns itself with levels that pose a risk to humans and disregards the welfare of the rest of the ecosystem.

Scientists have shown that acid rain kills aquatic insects and small soil organisms such as snails and grubs, which are critical foodstuffs for ducks and other birds. Toxic pollutants and agricultural chemicals that washed into water have ruined key feeding stops for migrating waterfowl in many areas, including Saginaw Bay. The U.S. Fish and Wildlife Service estimated that only 66 million ducks migrated during the fall of 1988, about 8 million fewer than the previous fall.

Wetlands aren't the only type of land that we've taken from plants, animals, birds and fish. Our subdivisions, resorts, office buildings and highways have shoved wildlife onto smaller and smaller areas of land. Wolverines and wolves need a lot of space to survive and they have run out of it. There only are a few wolves left in Michigan. The wolverines are gone.

Destruction of wildlife habitat by human activity is the main reason the passenger pigeon and the green trillium have vanished from Michigan. Other endangered species include the piping plover, the peregrine falcon, the whorled pegonia orchid, the copper belly water snake and the timber wolf.

Wildlife, plants and birds in Michigan are protected by the federal Endangered Species Act of 1973 and the 1974 Michigan Endangered Species Act. The laws call for a federal and state list of all animals, plants, insects and marine life that are threatened, endangered or extinct in the United States and in Michigan. Currently, the Michigan list, which is updated biannually, contains about 100 animals and more than 200 plants. Both acts forbid any human activities that would put any protected species in jeopardy or modify or destroy its habitat. Anyone who tries to harm or kill a protected animal or plant or import or export threatened and endangered species in Michigan is subject to fines of more than $1,500 per animal and 90 days in jail under the Michigan Wildlife Conservation Act. These acts are enforced by DNR conservation officers.

Researchers who wish to study, collect or conduct experiments on Michigan's various protected animal and plant species must apply to the DNR's Natural Heritage Program for a special permit. To encourage such work, grants of up to $4,000 are available through the Michigan Nongame Wildlife Fund, which is fueled through income tax checkoffs and donations to the DNR Natural Heritage Program, which oversees the fund. The Nongame Wildlife Fund receives about $500,000 each year from approximately 90,000 taxpayers. The money also is spent on developing plans to recover threatened or endangered animals and plants and to provide educational materials.

The fund has devoted a lot of effort toward helping the loon. Common loons, whose haunting calls were featured in the film "On Golden Pond," are threatened water birds that inhabit marshes and lakes in northern Michigan. These striking black-and-white birds are extremely sensitive to human disturbances while they are nesting and raising their young. There are only about 300 pairs of loons left in the state, a crisis for a bird whose female lays only two eggs during the nesting season.

Local citizens and DNR Nongame Wildlife Program officials have teamed up in an effort to revitalize Michigan's loon population. The Michigan Audubon Society backs a state Loon Preservation Association, which has coordinated a Loon Watch Program since 1985. The program enlists volunteers who live on northern lakes to keep track of loons' activities and educate their neighbors about the importance of staying away from nesting sites. Various lake-owner associations also have helped the DNR put up protective buoys and warning signs around sensitive loon breeding and nesting areas. Some places even have "loon rangers" who circle the lake and stop anyone agitating nesting loons.

Another bird that is in trouble in the piping plover. It is a black-and-white water bird with an orange beak that makes its home in wetlands along the Great Lakes shoreline, particularly along the eastern Upper Peninsula and northern Lake Michigan. The piping plover is listed as an endangered species on both the federal and state lists and its numbers are falling fast. Only 16 pairs of piping plovers are in Michigan, compared to 31 pairs in 1979. To protect piping plover chicks and eggs from pillaging attacks by marauding crows and gulls, state wildlife staff successfully erected cages around the nests. The DNR also closes the area around piping plover nests to public access during the nesting season in some parts of the state.

Another species drawing attention is the timber wolf. Although their image has been tarnished by fairy tales such as Little Red Riding Hood and The Three Little Pigs, wolves need humans' help in Michigan. Because they are predators, wolves serve a vital function of keeping other animal populations in balance in the northern ecosystem. But without public support, their population cannot be renewed. For years, farmers and hunters killed wolves because they believed they killed too many deer and livestock.

The densest population of wolves in the state lives on Isle Royale in Lake Superior, but within the past few years the number of wolves on Isle Royale has dropped dramatically from 50 to about 13. Researchers found that many wolves were infected with Lyme disease and canine parvovirus. The wolves may have been infected when campers brought their dogs onto the island illegally, or perhaps when they unknowingly carried the harmful virus or bacteria on their shoes or clothing. The wolves probably also were weakened genetically through years of inbreeding.

Another endangered species in Michigan, the Kirkland's warbler, can attribute its decline to firefighting improvements. Plump, green and yellow Kirtland's warblers will nest only in stands of young jack pines that are five feet to 20 feet high and six to eight years old. But jack pines only reseed themselves after their cones are exposed to fire. Presently, the Kirkland's warbler exists only in five northern Michigan counties.

Unless humans continuously plant young jack pines, the warblers will not be able to breed and their numbers will decline. The bird's habitat also is degraded when pines are harvested for commercial use or firewood or cleared for construction without any replanting program.

Because Kirtland's warblers are unique to Michigan, the DNR Nongame Wildlife Program has devoted a lot of effort to enhancing their population. In 1989, the DNR planted 1.8 million trees for the warblers on more than 1,500 acres. The U.S. Forest Service planted 1.3 million trees on 1,200 acres. These efforts appear to be paying off. In 1990, 265 nesting pairs of Kirtland's warblers were sighted, an encouraging increase from the 201 pairs counted in 1971.

The Karner blue butterfly also is an endangered species in Michigan which requires fire in its habitat. It is found in western Lower Michigan, particularly in the Allegan State Game Area. And while it has been sighted in seven other states, Michigan is thought to have the highest numbers of the butterfly. The Karner blue likes to feed on lupine flowers in dry, semi-open oak areas. The butterfly first developed in prairie grasslands that were periodically burned by grass fires, which prevented grasses and brushes from growing too high. Unless its habitat is maintained by humans with periodic prescribed burns, the Karner blue will not have the right conditions to live. Biologists from Huron-Manistee National Forest, the Nature Conservancy and the DNR are developing a management plan for the butterfly aimed at increasing its population.

The DNR also is planning to restore to Michigan the trumpeter swan, the largest species of waterfowl in the world. Trumpeter swans, with their distinctive all-black bills and extremely long necks, have been listed as extinct in Michigan but still exist in Alaska.

To try to restore the birds, the DNR acquired 20 trumpeter swan eggs from Alaska in 1989. The chicks are being raised at the Kellogg Bird Sanctuary near Kalamazoo and will be released when they are two years old. Then, hopefully, they'll make their home in marshes and lay their eggs in nests fashioned from abandoned beaver lodges and live again in Michigan.

Efforts to save the bald eagle also are beginning to pay off. Some of the Michigan bald eagles that were unable to reproduce after feeding on fish contaminated with DDT are showing signs of revival. In 1989, 165 bald eagle nesting pairs were sighted, compared to 80 pairs in the 1970s. Notably, more eggs were hatched in eagle nests that were located in inland areas because toxic contamination is still taking its toll on eagles who live close to the Great Lakes shoreline. Only half of the eggs survive when hatched along the shore.

But as population continues to grow in the coming years, more humans will be infringing on natural habitats to build their homes, offices, golf courses and shopping centers. With diminishing food, water, space and protective cover — the key ingredients of a habitat — even the hardiest species of wildlife will be hard-pressed to survive. Because our Great Lakes ecosystem is so complex, if one species is lost many others will surely follow.

Aldo Leopold, considered to be the father of conservation, once said the key to intelligent tinkering is to keep all of the parts. We have been carelessly tinkering with wildlife and have been losing some of the parts. We have managed to reverse, in just a few generations, much of nature's craftsmanship, which took ages to create. The human stresses on Michigan's wildlife populations will only become more severe. We will have to work harder to maintain open spaces for wildlife, educate ourselves about how our actions affect the plants and animals near us, and try to live in a way that can accommodate all members of the ecosystem.

SOLUTIONS

Things to do:

• Don't let helium balloons go. Even from the Midwest they can end up in the ocean and kill animals.

• Snip your six-pack rings so they don't end up around the neck of a water bird or fish.

• Litter ruins habitats, especially plastic and foam. Recycle it or throw it in a trash barrel.

• Plant shrubs and trees that provide a variety of food such as berries, seeds, nuts or nectar.

• Add a bird bath, shallow pond or stream to your yard as a source of water. It will attract many different species of wildlife.

• Put up a bird feeding station with a variety of bird feeders and seed to attract many different kinds of birds, from hummingbirds to large hawks.

• Build bird houses to provide shelter and nesting areas. Set out balls of string, strips of paper and yarn for nesting materials.

• Make bat houses to attract bats to your yard. They are a great natural insect control.

• If you live in an apartment, put up window boxes and feeders for birds.

• Adopt a stream or stock a pond with fish. Contact your DNR district fisheries biologist to find out which species would best live there.

• Contribute to the Michigan Nongame Wildlife Fund on your state income tax form or make a direct donation to help protect Michigan endangered and threatened species.

Things to do for your community:

• Join a beach or river cleanup. Call your local nature center or parks system for information.

• Keep track of proposed changes for natural habitats in your area. Contact your city or county planning department for information.

• Help a local conservation group that is working to save the natural area. Get involved with your local nature center or wildlife reserve.

• Teach children to appreciate and protect the plants and wild animals in their environment.

Contacts:

• Contact the Michigan Natural Heritage Program and Nongame Wildlife Program to find out what rare or endangered species are found in your area and what you can do to help protect them. Michigan Natural Heritage Program, Nongame Wildlife Program, Department of Natural Resources, Box 30028, Lansing, Mich. 48909.

• If you witness a violation of the Endangered Species Act or state fish and game laws, call the DNR at 1-800-292-7800.

• To report violations of federal fish and game laws, call 1-313-941-3388.

• For information about Michigan's wetlands, contact the Department of Natural Resources Land and Water Management Division, Box 30028, Lansing, Mich. 48909, (517) 373-1170. Request the division's free brochure, "Michigan Wetlands: A Guide for Property Owners and Home Builders."

• Another source of wetlands information is the U.S. Army Corps of Engineers, Detroit District, Box 1027, Detroit, Mich. 48231, (313) 226-2218.

• Other DNR resources: Wildlife Division, 6th Floor, Mason Bldg., Box 30028, Lansing, Mich. 48909, (517) 373-1263. Fisheries Division, 4th Floor, Mason Bldg., Box 30028, Lansing, (517) 373-1280. Forest Management Division, 5th Floor, Mason Bldg., Box 30028, Lansing, (517) 373-1275.

• For information about Michigan's two national wildlife refuges, the Seney and the Shiawassee, contact the U.S. Fish and Wildlife State Office at 1405 East Harrison Road, East Lansing, Mich. 48823, (517) 337-6652.

• The Michigan United Conservation Clubs' "Wetlands Watch" can notify you about wetlands alterations in your community. For information about attracting wildlife to your backyard, ask for "Conservation Planting," "Invite Wildlife to Your Backyard," and "Invite Birds to Your Home." Also ask for plans to build birdhouses. There is a $1 postage and handling fee. MUCC, Box 30235, Lansing, Mich. 48909, (517) 371-1041 or 1-800-777-6720.

• For information about preserving Michigan wetlands, consult "Michigan Wetlands, Yours to Protect: A Citizen's Guide to Local Involvement in Wetlands Protection." Published by the Tip of the Mitt Watershed Council, Box 300, Conway, Mich. 49722.

• The Izaak Walton League sponsors a stream adoption program,

Save Our Streams Program. 1401 Wilson Blvd., Level B, Arlington, Va. 22209, (703) 528-1818.

• To find out how to build bat homes, send $1 to Bat Conservation International, c/o Pat Morton, Box 162603, Austin, Texas 78716.

• The National Wildlife Federation provides a brochure on how to turn your backyard or school grounds into a wildlife habitat. NWF Backyard Wildlife Habitat Program, 8925 Leesburg Pike, Vienna, Va. 22184-0001.

• The Nature Conservancy buys land to protect it. Michigan Chapter is located at 2840 East Grand River Ave., East Lansing, Mich. 48823. (517) 332-1741.

• The Michigan Audubon Society, 409 West E Ave., Kalamazoo, Mich. 49007, (616) 344-8648; or the Detroit Audubon Society, 121 South Main St., Royal Oak, Mich. 48067, (313) 545-2929.

• Michigan Loon Protection Association, 409 E Ave., Kalamazoo, Mich. 49007.

• Michigan Nature Association, 124 Miller, Mount Clemens, Mich. 48043, (313) 468-1313.

• Michigan Wildlife Habitat Foundation, 6425 South Pennsylvania, Lansing, Mich., (517) 882-3110.

• Michigan Lakes and Streams Association, Don Wynne, 11262 Oak Ave., Three Rivers, Mich. 49093, (616) 244-5336.

• Michigan Natural Areas Council, University of Michigan Botanical Gardens, 1800 North Dixboro Road, Ann Arbor, Mich. 48105, (313) 278-9269.

• For information about turning abandoned railroad tracks into natural wildlife refuges and recreation areas, contact Rails to Trails Conservancy, Michigan Chapter, Box 23032, Lansing, Mich. 48909.

• Wildflower Association of Michigan, Box 4460, East Lansing, Mich. 48826-4460.

• Ducks Unlimited, National Office, One Waterfowl Way, Long Grove, Ill. 60047, (312) 438-4300.

- Pheasants Forever, Box 75473, St. Paul, Minn. 55175, (612) 481-7142.

- Ruffed Grouse Society, 1400 Lee Drive, Corapolis, Penn. 15108, (412) 262-4044.

- Trout Unlimited, National Headquarters, 501 Church St. NE, Vienna, Va. 22180.

- Wildlife Management Institute, Suite 725, 1101 14th St. NW, Washington, D.C. 20005, (202) 371-1808.

- World Wildlife Fund, U.S. 1255 23rd St. NW, Washington, D.C. 20037, (202) 293-4800.

- Huron-Manistee National Forests, 421 S. Mitchell St., Cadillac, MI 49601 (800) 999-7677

Further reading:

Wetlands, The Audubon Society Nature Guides, William A. Niering, Alfred A. Knopf Publishing, New York, January 1987.

A Sand County Almanac, by Aldo Leopold, London: Oxford University Press, 1989.

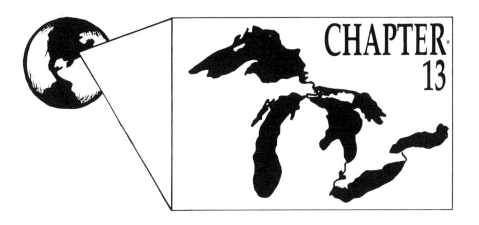

Are We Destroying Our Chance For Survival?

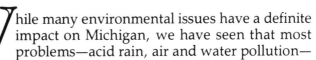
While many environmental issues have a definite impact on Michigan, we have seen that most problems—acid rain, air and water pollution—are regional, national and global in their scope. It is not only important for us to defend our state from environmental degradation but also to work with other states and nations to tackle the problems that effect every organism on this planet. For international environmental threats of ozone depletion, climate change, overpopulation and deforestation, the old motto "think globally, act locally" is now more appropriate than ever.

If the earth were an apple, the ozone layer would only be about as thick as the skin. But while ozone molecules are relatively scarce compared with other components of our planet's atmosphere, without them the sun's destructive ultraviolet rays would be free to convert the earth into a desolate, inhospitable planet devoid of wildlife or vegetation as we know it. Some scientists say that unless we do something soon, that may happen.

Experiments have demonstrated that chlorofluorocarbons (CFCs), a group of synthetic chemicals often known by the trade name Freon, are causing ozone to break up in the stratosphere 10 to 30 miles overhead, leaving us vulnerable to harmful radiation from the sun. CFCs have been used for many things since they were discovered in the 1950s. They propelled whipped cream, deodorant and hair spray from aerosol cans, acted as a coolant in refrigerators and air conditioners and formed foam containers for eggs and hamburger. Whenever products made of CFCs are manufactured or thrown out they eventually release CFC molecules into the air. If they are burned in an incinerator plastic materials will emit chlorine particles which form into hydrogen-chloride, a hazardous acid gas that causes respiratory problems, harms plants and corrodes cars.

Both plastic and foam are made from chemicals which are produced when crude oil is refined into other petroleum products such as gasoline. Many of those chemical raw materials are toxic. According to the Environmental Protection Agency, five of the top six chemical production processes that generate the most hazardous wastes are used to make plastic packaging.

Two of these by-products, ethylene and benzene, form ethylbenzene, the main ingredient for making polystyrene. Once a sheet of polystyrene is formed, the manufacturers inject it with gas which expands when exposed to heat, forming millions of gas-filled bubbles. Usually the gas escapes during or shortly after the manufacturing process, leaving behind air-filled beads that can be melted and shaped into foam products such as cups or plates.

For years those insulating bubbles were created by chlorofluorocarbons, a chemical that destroys ozone molecules. When the hole in the ozone layer was discovered over the South Pole in 1985, the foam industry learned that it had released enough CFCs into the atmosphere to account for a sizeable portion of known ozone destruction. In an undisturbed state, the ozone layer absorbs 99 percent of the sun's ultraviolet radiation. Three years later, the industry promised to stop using CFCs to make food service products by 1989. While this is a step in the right direction, other foam manufacturers who make commercial packing materials, cool-chests and other products still use CFCs.

Once they get into the atmosphere, these extremely stable compounds will remain there for more than 100 years until they gradually make their way into the higher stratosphere.

This is where the problems begin. The only thing that will destroy a CFC molecule is the intense heat emanating from ultraviolet rays in the upper stratosphere. Ultraviolet radiation breaks CFCs into their basic components: carbon, fluorine and highly reactive chlorine. However, ozone molecules, which consist of three oxygen atoms, are extremely unstable. Errant chlorine atoms attack the ozone and rob it of an oxygen atom to form chlorine monoxide, leaving a pure molecule of oxygen (two oxygen atoms) to go free. Usually, the single atom of oxygen in the chlorine monoxide molecule will bond with another free atom of oxygen, which leaves the chlorine atom ready to destroy more ozone molecules. It has been estimated that one atom of chlorine that is released into the stratosphere can destroy 100,000 molecules of ozone during the 100-year period that it will remain there as a catalyst.

Scientists believe that the use of CFCs all over the world is probably responsible for the alarming reduction in the amount of ozone in the stratosphere today. An estimated two percent of the ozone layer already has been depleted in most parts of the world, but the actual amount of depletion is difficult to measure because ozone levels fluctuate from year to year by two to four percent. Based on the amount of CFCs that have been released and are being manufactured worldwide, NASA predicts a 10 percent depletion that will continue for nearly a century. CFCs released into the atmosphere since 1955 have not yet reached the stratosphere, where they will accelerate ozone deterioration.

A problem-free solution is not forthcoming. One of the substitute gases used to make foam for food containers, pentane, does not destroy ozone but does contribute to urban smog. The other replacement gas, called HCFC or CFC22, has the potential to destroy 95 percent less ozone than the previously used CFCs but still poses some threat to the ozone layer.

What will happen if our protective ozone layer is significantly deteriorated? We won't really know until it happens, but scientists have made some educated guesses. Anyone who has been badly sunburned can probably imagine what life would be like without the ozone layer. Experts are predicting a much higher incidence of sunburn and, consequently, more cases of skin cancer. Increased exposure to ultraviolet radiation also can cause cataracts and weaken our immune systems.

Additional ultraviolet rays would stimulate the formation of more ground-level ozone and urban photochemical smog, and coupled with the already damaging effects of acid rain and smog, would be harmful to agricultural cropland, forests, fish and wildlife. Ozone depletion also may heighten the world's acid rain problem by adding more hydrogen peroxide into the atmosphere, which accelerates the transformation of sulfur dioxide and nitrogen oxide into sulfuric and nitric acids.

In an effort to stop this frightening destruction, the governments of several countries have taken a few preventative steps. In 1978, the United States, Canada and most Scandinavian countries decided to ban the use of CFCs in aerosol spray cans. CFCs still are used in refrigerators, air conditioners and some plastic foams. In 1987, 24 countries, including the United States, signed the Montreal Protocols, an agreement to cut CFC production in half over 10 years. By 1989, 66 counties had joined the pact.

We still have a long way to go to achieve the international cooperation that is imperative to reverse the destruction we already have unleashed on the ozone layer. Sixty-six countries take up only a small portion of the world map. Too many nations are selfishly concerned only with short-term economic profits rather than overall long-term survival. Advanced industrial countries, like the United States and Great Britain, need to educate other nations about alternatives to CFCs and the importance of maintaining ozone in the stratosphere. If voluntary cooperation cannot be achieved on a global scale, the United Nations might begin to enforce vital international environmental agreements such as the Montreal Protocols.

The world faces other threats. When working properly, the greenhouse effect is a necessary, life-sustaining function of the earth's atmosphere. Certain naturally occurring gases such as carbon dioxide (CO_2) and methane form a blanket in the atmosphere, which allows the sun's rays to enter but traps their heat to warm the planet. If greenhouse gases like CO_2 did not trap the sun's energy, the earth would be an icebound planet like Mars.

The amount of solar heat that is allowed to enter and escape the earth's atmosphere has been kept in balance for thousands of years by natural greenhouse gases. But many scientists believe that human activity has interfered with this process, accelerating the warming by adding an excess of greenhouse gases into the atmosphere.

Fifty percent of the greenhouse effect is believed to be caused when CO_2 is emitted into the air, primarily from power plants, car exhaust and burning coal, oil, gasoline and natural gas. Whenever large stands of trees—such as tropical rain forests—are cleared for development, more CO_2 is released into the atmosphere. Methane gas, nitrogen oxide, CFCs and ground-level ozone collectively contribute to the rest of the greenhouse effect.

Methane gas is a by-product of natural decay and is prevelant around municipal solid waste landfills and livestock manure. It also is emitted by rice paddies and termites in the process of digestion. A single termite can generate five liters of methane gas per minute. The next contributor to the greenhouse effect, nitrogen oxide, mostly is produced by gasoline combustion in motor vehicle engines but also occurs when chemical fertilizers break down. CFCs have been banned for use as propellants for spray can products, but still are used in air conditioning and refrigerators. Ground-level ozone is caused by automobile and industrial emissions.

What can we expect if the protective blanket of gases becomes so thick that fewer solar rays can escape and too much heat remains trapped near the earth's surface? The extremely hot and dry summer of 1988 may be a harbinger of what is to come. Five of the hottest years in the last century have occurred during the 1980s. Although there is no conclusive evidence that this was caused by the greenhouse effect, many scientists believe it is very likely. Since the end of the last ice age 18,000 years ago, average global temperatures have not increased more than 3.6 degrees Fahrenheit. But recently, several studies have predicted that over the next 100 years, average global temperatures could rise 4.5 to 9.9 degrees Fahrenheit.

The studies used computer models to try to predict changes the world would undergo if temperatures increase as much as predicted. The picture they paint is grim. Climate changes could cause severe and dramatic shifts in weather patterns and a greater incidence of tropical storms. Hotter weather could cause more frequent droughts, gradually converting the Midwest farm belt and other fertile areas into desert lands. This would sharply reduce the amount of food that is available to feed an ever-expanding world population.

Because water expands when it is heated, the oceans would rise if global warming occurs. Higher temperatures threaten to melt mountain glaciers and the polar ice caps, adding more water to the world's oceans. Studies show that a seven percent increase in average global temperatures would cause sea level to rise two feet. Coasts along India, China and the United States would be underwater. Drinking water supplies would be contaminated with saltwater.

While these predictions create a frightening, gloomy vision of the future, they are still surrounded with scientific uncertainty and controversy. Still, enough evidence exists to suggest that we would be wise to quit loading the atmosphere with greenhouse gases.

The first step could be to cut back on burning carbon-producing fossil fuels, the largest contributor to global warming. Before information about global warming was widely available, most energy analysts predicted that we would not turn to renewable, alternative energy sources like wind, solar and geothermal power until we had exhausted the last of our fossil fuel resources. But now, several studies have indicated that if we wait until fossil fuels run out before switching to cleaner energy sources, it may be too late. Our past actions already may have set in motion the destruction we fear.

One of the simplest, least expensive methods of cutting global CO_2 production is to conserve energy. Currently, the world produces 6 billion tons of CO_2 per year by burning fossil fuels for energy. But industries can maximize their energy efficiency by using co-generation, a process that captures the intense heat and steam created during production and use it to produce electricity. It can run the plant or be sold to a utility company. Home owners can invest in better insulation and more efficient light bulbs and appliances to save money and energy. Cities could switch to fuel-saving forms of public transportation, car pooling and bike lanes. The federal government could mandate improved automobile gas mileage.

Nuclear power plants, unlike coal-fired facilities, are a clean-burning source of energy for residential heating and electricity. They do not produce CO_2 or nitrogen oxide emissions, but they do produce dangerous radioactive wastes, which are virtually impossible to store safely on a long-term basis. This storage dilemma, along with the catastrophic nuclear accident at Chernobyl and the near-disaster at Three Mile Island, have prejudiced public opinion against this type of power. As a result, very few nuclear plants are being constructed today.

Harnessing wind to create power is another alternative to burning carbon-based fuels. It is an unlimited energy source that does not create any air emissions or produce hazardous wastes. However, it is only effective in places with strong, consistent winds. Rows of wind turbines, called a wind farm, are being used in windy parts of California to generate 90 percent of all U.S. wind power. California is an ideal place for wind power because its windy places are near utilities that can convert it into electricity. Other places that would be suitable for wind power are too remote to be useful. In Denmark, floating wind farms catch strong offshore winds to generate electricity that is relayed back to shore.

Another natural method of producing energy is to capture the geothermal heat generated from hot springs, geysers and underground steam and hot water deposits. It can provide residential or industrial heat and electricity. Approximately 65 percent of all homes in Iceland depend on geothermal energy for heating. Its drawback is that it creates moderate air pollution with emissions of hydrogen sulfide, ammonia and radioactive substances. It may cause water pollution from underground deposits of mercury and boron and it also is not a resource renewable in a single lifetime. Each geothermal deposit contains a supply of energy that will last about 100 to 200 years.

Solar power appears to be the most promising energy alternative. The sun's rays are an unlimited raw material which can be harnessed to heat buildings, water and provide electricity. Even though the sun is 93 million miles away, it can provide enough heat and light in three days to equal the amount of energy that would be produced by burning all of the planet's oil, coal and timber reserves. A solar panel the size of Lake Erie would collect enough power in one day to meet the energy needs of the entire United States for one year.

The sun's energy can be converted into power through either passive or active systems. Passive solar systems absorb the sun's rays and turn them into heat directly inside specially designed buildings. Solar structures incorporate large walls of south-facing windows and interior walls and floors covered with absorbent brick or tile. A black solar collector, containing water or chemicals, absorbs the sun's heat and slowly releases it to warm the building. Window space is kept to a minimum on the north side of the structure to prevent heat loss.

Active solar systems contain solar collectors that are designed to concentrate the sun's energy. The absorbed solar energy is pumped into a storage site where it can be distributed to heating ducts or water heaters as needed. Approximately 50 percent of all of the homes in Israel use solar energy to heat their water. And by 1995, Japan expects to have solar collectors on 30 percent of all of its buildings.

Solar energy also can be used to produce electricity. Photovoltaic cells which use superconducting materials for collectors are the most efficient way to convert sunlight into electricity. Silicon wafers generate a spark of electrical current when they are hit by the sun's rays. When enough sparks are collected from a bank of cells, they can generate electricity for residential use. Currently, the use of solar power for heating is more widespread because the equipment for transforming solar energy into electricity is still quite costly.

Transportation is another area in which we can conserve energy. Automobiles are the largest source of nitrogen oxide and carbon dioxide and are the second largest source of CO_2 in the United States, surpassed only by electrical power plants. The average American car driven approximately 10,000 miles per year releases its own weight in carbon into the atmosphere each year. Americans alone account for about one-tenth of the world's commercial oil consumption by driving back and forth to work.

To fight the threat of global warming, automobile-based countries like the United States will have to become less dependent on cars and trucks as their primary mode of transportation. Cheaper, more convenient public transportation, improved fuel economy and the use of alternative fuels could greatly diminish the amount of greenhouse gases emitted. The fleets of trucks crossing the highways also will need to be scaled down and incorporated into a more efficient freight hauling scheme.

Studies have shown that raising standards for fuel economy in the United States by just five miles per gallon would eliminate 100 million tons of carbon dioxide per year. Another way to rid the atmosphere of carbon dioxide is to power vehicles with alternative fuels that do not emit carbon as a by-product of combustion. For example, 62 percent of all automobile fuel that is sold in Brazil is ethanol, a gasoline alternative derived from sugar cane. Brazil's alcohol fuel program is the world's largest, with 72 million barrels of ethanol produced annually. The United States is the second largest producer, generating 20 million barrels per year of a corn-based fuel, where it is mixed with regular gasoline.

In countries that are too poor or too populated to be automobile-centered, bicycles are people's primary means of transport. In China and Taiwan, the streets are clogged with bicycles used for hauling goods and passengers. In the Netherlands, 80 percent of the people who commute to work by train get to the station on a bicycle.

There are approximately 800 million bicycles in the world and twice as many people ride bikes as drive cars. Bikes can eliminate urban traffic jams in which thousands of idling cars send up a cloud of CO_2 and nitrogen oxide, the precursors to global warming, acid rain, urban smog and health problems. Combining mass transit and bicycling would greatly reduce traffic jams and the pollution they create.

But all our efforts to improve our climate and surroundings will be for naught if we use up all of our resources. Stabilizing world population is as important as controlling global temperatures. Acid rain, air pollution, global warming and water contamination are all aggravated by overpopulation. According to a report by the United Nations' population committee, the world's population grows by three people every second. The Population Crisis Committee, a Washington, D.C., think-tank, estimates that world population now stands at 5.3 billion and will double in 39 years. This will double global food, water and fossil fuel consumption.

If the number of mouths keep growing, there won't be enough food to feed them. Increasing population places huge demands on underground aquifers, which can turn to saltwater if they are depleted more rapidly than nature can replenish them. An unchecked population will need more energy for housing and lighting, further contributing to global warming.

There are two types of overpopulation. The first occurs mostly in less developed areas, where there are too many people trying to squeeze enough food, water and fuel out of a limited amount of resources. The second type is better described as over consumption. In highly industrialized nations, fewer people furiously consume resources, producing more pollution and environmental destruction per capita than in a less developed, but more populated country. Industrialized nations represent less than 25 percent of the world's population but consume 75 percent of its energy resources.

Population is difficult to control because of the many social, cultural and religious viewpoints that touch upon it. But reducing the number of people that the earth supports is not an impossible undertaking. Countries such as China and Japan, which were stressed by a swelling citizenry, cut their national population growth rate in half in six to seven years. Although some have objected to the manner in which these results were achieved—the governments allegedly pressured women to have abortions or undergo sterilization after the birth of their first child—the fact remains that population control is essential. Making birth control available to all who wish to use it is a reasonable step.

It is equally important to educate people about the relationship between family size and quality of life. As George Bush wrote in the introduction to a 1973 book called World Population Crisis, "Success in the population field may determine whether we can resolve successfully the other great questions of peace, prosperity and individual rights."

As for the other type of overpopulation — over consumption — affluent countries have to reduce their appetite for natural resources and material goods. It will be a shocking behavorial change, but one that must occur.

As President Franklin D. Roosevelt said during one of his fireside chats in 1942, ''Never before have we had so little time in which to do so much.''

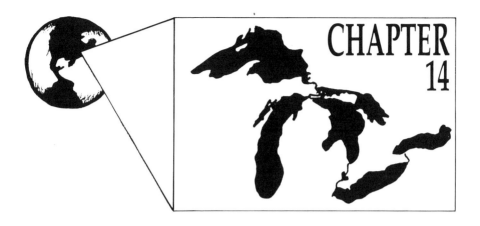

Forests Can Reverse The Damage

Luckily, there is a simple way to begin to correct the damage done to air quality. Plant trees. Trees fight global warming and purify the air by absorbing CO_2 and other air pollutants and releasing oxygen. But if they are cut down, they not only quit absorbing CO_2 but release it into the atmosphere, resulting in a double threat to the stability of the planet's climate. Deforestation results in an estimated 10 percent of the world's total CO_2 production.

Trees provide a variety of important functions. Tree roots prevent erosion and trees provide cool shade in urban areas. Trees can be planted on pieces of land that are not suitable for farming or buildings to offset the effects of deforestation or CO_2 emissions from power plants.

In 1988, the American Forestry Association formed an organization called Global Releaf which intends to encourage communities to plant 100 million trees by 1992. If successful, the plantings would absorb an estimated 5 million tons of carbon per year. Although this amount may seem small in comparison to the daunting 6 billion tons of carbon dioxide that is produced every year globally, planting trees still plays a valuable role in the world's efforts to stabilize its climate. It gives people a tangible, symbolic act against the degradation of the planet and encourages good stewardship of the land.

Forests play a vital role in regulating the planet's environmental cycles. Trees filter carbon out of the atmosphere during their growth process and replenish the air with oxygen. They also regulate the earth's flow of water, especially in areas that receive frequent, heavy rains. During the rainy season, forests act like a sponge to soak up excess water, protecting nearby inhabitants from floods and property damage. When things begin to dry up, the trees slowly release the stored water to keep the soil moist.

Forests are important for maintaining soil composition and nutrients. Their roots cling to the soil, preventing soil erosion and when their leaves and branches fall to the ground and decay, they enrich the soil as fertilizer. Forests also support a wide variety of wildlife.

When trees are chopped down in the name of human progress, the forest's natural cycles are disrupted and the environment suffers. When large areas of trees are cleared, the forest's sponge effect no longer functions and dangerous floods advance unchecked. Rainfall runoff in India is causing severe flooding as it slides down bare hills that were once covered with trees. In Bangladesh, flooding induced by deforestation is destroying crops. Without tree roots to anchor the soil and keep it moist, it blows away in the wind. Overgrazing cattle in areas that were once forests can become barren, desert-like land.

Before it was settled by Europeans, the United States was covered with lush forests. But since the first colonists arrived in 1607, 45 percent of the country's forests have been cut down. Because of poor forest management techniques years ago, the United States has very few old-growth forests left. The last remaining stands of old-growth trees are located in the Pacific Northwest, where conservationists are trying to protect them from logging companies. These old-growth forests support many species of wildlife as well as the spotted owl, which will become extinct if its habitat is destroyed. Since better forestry practices were put into use in the 1920s, the nation's forests have remained stable, occupying about one-third of the country's total surface area.

But the world's most important forests are the rain forests and they are in danger. Most of the world's tropical rain and evergreen forests are located on a band around the equator in Central and South America, Africa and Asia. Although they only cover between two and seven percent of the earth's surface, tropical rain forests contain more than half of the world's known species of plants, animals and insects.

Trees are so thick in a rain forest that only two percent of the sun's rays successfully penetrate the thick overhead canopy of leaves and branches. Although little light reaches the floor of a rain forest, millions of species of plants, animals and insects thrive there because the climate is very stable. Tropical rain forests host tall, spindly, shallow-rooted trees, tangled vines, hanging epiphytes, toucans, tapirs, tropical fruits and the fruit bats that scatter their seeds.

But remote-sensing satellite photographs show that more than 40 percent of the world's original growth of tropical rain forest has been destroyed, mostly within the past 50 years. According to the World Resources Institute, every year a patch of tropical rain forest the size of the state of Washington is lost—a total of 40 to 50 million acres per year, based on the latest satellite information. Experts say the worst destruction is taking place in Brazil, Costa Rica, India and Indonesia.

As the world's population grows, large expanses of rain forests are being cleared to grow cash crops, raise cattle, harvest exotic woods and construct buildings, roads or power plants. In the past 20 years, nearly 40 percent of the rain forest in Central America has been converted to pasture for cattle-raising, 90 percent of which is exported to the United States for use in fast food restaurants.

The United States and Japan also import the lion's share of exotic tropical woods such as mahogany and teak for furniture.

Perhaps the most tragic environmental impact of tropical rain forest destruction is the loss of unique species. Every day, at least one tropical rain forest species becomes extinct because of disruption in its habitat, according to Friends of the Earth, a global environmental group. Eighty percent of the native flowering plants in Madagasgar do not grow anywhere else in the world. Half of its rain forests have been destroyed, putting at risk the rosy periwinkle which is used to make medicine that fights leukemia, Hodgkin's disease and other types of cancer. In fact, 70 percent of the plants that have been identified by the National Cancer Institute as containing cancer-fighting compounds only will grow in a tropical rain forest. Rain forest plants are used for other medicinal purposes including anesthetics, some types of birth control pills and tranquilizers. One-quarter of the prescription drugs used in the United States are derived from a plant that comes from the rain forest.

A few countries, such as Costa Rica, Haiti and the Philippines, are planting trees and turning what is left of their rain forests into wilderness preserves. Others, such as Bolivia, are negotiating debt-for-nature bargains with industrialized nations where they agree to protect their forests if their lenders promise to decrease their debt payments.

Most of the mammoth Amazon basin rain forest is located in Brazil. That country's economy is in trouble and suffers from sky-high inflation. As overpopulation continues, people clear the rain forests for agriculture, fuel and food. Brazil and other rain forest countries were given World Bank International Monetary Fund loans to eliminate this need, but the money has been mismanaged and the money isn't going to the people who need it most. Poverty leads to destruction of the rain forests.

Taking steps to reduce the potential threat of global warming has several benefits, whatever one's opinions of scientific predictions may be. Cutting emissions of CO_2, nitrogen oxide and CFCs will reduce acid rain, smog, ozone layer depletion and rain forest destruction. Other improvements include cheaper, cleaner and more efficient sources of energy and inexpensive, convenient public transportation. Reducing fossil fuel consumption also reduces overreliance on foreign sources of oil and the wars that come with it.

But getting every nation to decrease fossil fuel consumption to reduce the threat of global warming is a seemingly impossible task. While areas that face possible desertification or coastline destruction may be supportive of efforts to prevent global warming, areas that could benefit from a wetter, warmer climate will not. Change comes slowly even at the national level. Reconciling international differences involved in diminishing the threat of global warming is a formidable challenge.

SOLUTIONS
Things to do:

• Plant trees. Trees remove CO_2 from the air, lessening the threat of global warming.

• Switch to energy-efficient fluorescent light bulbs and buy the most energy-efficient appliances, homes and cars on the market. Consider new products in terms of lifetime rather than short-term cost. Use natural lighting instead of electric lights whenever possible.

• Hang your clothes outside to dry whenever possible.

• Get heat and cooling from natural sources as much as possible.

• If you're buying a new home consider a solar or energy efficient design.

• Plant deciduous trees on the south side of your house to provide shade in the summer. They will let sunlight through in the winter once their leaves fall.

• Install overhead fans or use portable fans rather than air conditioners, which use more energy and emit CFCs. Keep rooms cool with shades.

• Insulate your exterior walls and attic.

• Lower your thermostat 10 degrees to 15 degrees each night in the winter.

• Purchase a "setback" thermostat.

• Buy an energy-efficient water heater—up to 40 percent of your utility bill is spent for heating water. Wrap insulation around your water heater.

• Wash clothes in cold water as much as possible.

• Do dishes by hand more often and only run the dishwasher when it's full.

- Keep your car tuned for maximum fuel efficiency.

- Don't buy products such as some types of insulation and certain aerosol sprays that contain ozone-depleting CFCs.

- Avoid purchasing furniture and other products made of tropical hardwood such as teak and mahogany.

- Exotic pets such as cockatoos, boa constrictors and iguanas often are taken illegally from tropical rain forests. Find a dealer you can trust and only buy ones that were captive bred.

Things to do for your community:

- Join a bike club, encourage your city government to create bike lanes and make the city more bicycle user-friendly. Fewer car drivers means fewer exhaust emissions.

- Launch a tree-planting campaign in your neighborhood, business or school.

- Help an organization like the Rain Forest Action Network that is working to preserve the world's rain forests.

- Support strong legislation against the production and use of CFCs. Urge federal officials to work toward international agreements to stop the depletion of the ozone layer and slow global warming.

- Don't buy products made from tropical hardwoods and pulp.

Contacts:

- The American Forestry Association's Global Releaf Program, Box 2000, Washington, D.C. 20013, or call (202) 667-3300.

- For a copy of the brochure "A Consumer's Guide to Protecting the Ozone," send $1.50 to the National Toxics Campaign, 29 Temple Place, 5th Floor, Boston, Mass. 02111.

- The Rain Forest Action Movement in Ann Arbor runs educational and lobbying projects to raise forest awareness. Contact them at 430 East University, Ann Arbor, Mich. 48109, (313) 662-0232.

- U.S. Department of Energy's toll-free number provides information and referral for questions about energy conservation and renewable energy sources: 1-800-523-2929.

Contacts:

• Call the National Wildlife Federation's Legislative Hot Line for weekly updates on federal environmental legislation: (202) 797-6655.

• If you have questions about federal environmental regulations but do not know which agency to contact, call Michigan's Federal Information Center: Detroit, (313) 226-7016 or Grand Rapids, (616) 451-2628.

• By calling the congressional Legislative Status phone number, (202) 225-1772, you can find out the status of pending environmental legislation in Congress.

• The Capitol Switchboard, (202) 224-3121, can provide general information and can connect you to your U.S. senator or representative.

• For help finding the right agency to answer your environmental questions, contact the Michigan Department of Natural Resources Information Services Center: (517) 373-1220.

Further Reading:

• "Making Congress Work for Our Environment," a brochure published by the National Wildlife Federation that outlines the legislative process, gives guidelines on how to express your views to elected officials. NWF also prints "Use the News to Protect Your Environment: A Primer on the News Media." 1400 16th St. NW, Washington, D.C. 20036-2266.

50 Simple Things You Can Do To Save the Earth, by The Earthworks Group, Earthworks Press, 1400 Shattuck Ave., Berkeley, Calif. 94109. It provides tips to solve the world's environmental problems. Available in many bookstores.

Save Our Planet: 750 Everyday Ways You Can Help Clean Up the Earth, by Diane MacEachern. A handbook for environmentally concerned individuals.

State of the World 1991, the latest in a series about global environmental issues published each year by the Worldwatch Institute. Worldwatch also publishes a set of papers on various world environmental/economic problems. Contact the Worldwatch Institute at: 1776 Massachusetts Ave. NW, Washington, D.C. 20036, (202) 452-1999.

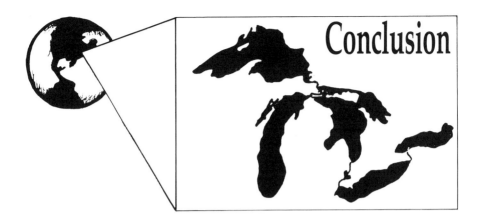

What Is A Consumer To Do?

Americans can play a critical role in dictating what to do with the natural resources we have left. We can use them up at a furious pace or try to preserve them. If we choose the latter, one of the first steps is to consume carefully. Businesses are not going to make that easy.

Many American corporations have sprouted green leaves, but are they real or artificial? This new wave of "green" products may leave consumers bewildered, confused and eventually cynical. A healthy dose of cynicism is probably wise in these days of "biodegradable" garbage bags and McGreen fast food chains. Many companies are spending more money on pages and pages of landfill-clogging advertisements about their "green" good deeds.

Environmental awareness, or at least a guise of it, is good business. For example, in response to the public outcry over the Exxon Valdez disaster, a coalition of environmental, religious and investment groups persuaded 3,000 companies to sign a 10-point environmental agenda called the "Valdez Principles." It was intended to serve as a measurement for comparing corporate environmental responsibility. The agreement calls upon companies to cut back their consumption of resources, cut waste and conduct an annual environmental audit on themselves. Ironically, Exxon refused to sign the principles because "they interfered with ordinary business," according to newspaper accounts.

The recent degradable plastic fiasco is a perfect example of the rampant consumer confusion that may cause newly-conscientious shoppers to throw in the towel.

For example, what is a newly-aware shopper supposed to do when he or she encounters degradable garbage bags? They sound irresistible. Resist them.

"Degradable" means that the manufacturer added cornstarch to the plastic so that it will fall apart more easily than regular plastic, but only in the presence of sunlight. A Greenpeace report, entitled "Breaking Down the Degradable Plastic Scam," indicates that some biodegradable plastic products may emit hazardous substances when (and if) they break into pieces.

Okay, so they fall apart. But they fall apart into tiny pieces of indestructible plastic. Regular bags remain large pieces of indestructible plastic. At least ordinary bags don't hide behind false claims of environmental-friendliness.

Although this barrage of "environmental" product promotion amid conflicting opinions and politics of environmental "experts" is bewildering, consumers owe it to themselves and to the environment to dig in and try to make some sense of it all. Without committed and vigilant consumer pressure, the recent "green" revolution in the marketplace is likely to fizzle out as quickly as it arose.

To help consumers distinguish the genuine environmentally responsible companies from the hucksters, a group called the Alliance for Social Responsibility, backed by Earth Day organizer Denis Hayes, has been working on a uniform standard for a "Green Seal" of approval similar to the one that is already being used in Canada and parts of Europe.

One of the most important things shoppers can do is to praise stores and companies that carry or manufacture products made from recycled materials. If recycling is ever going to succeed in this country, we have to find a market for the mountains of newspapers and milk jugs that have been piling up in recycling centers. Sometimes they are dumped back into the landfill in neatly separated mounds.

When you're in the store, look for the recycling symbol of three chasing arrows. If the symbol is filled in or says "recycled," it means it has been made from recycled materials. If the symbol is open or labeled "recyclable," the product has been made from virgin raw materials but can be recycled.

Reject products wrapped in packaging that clogs landfills, cannot be recycled, or result in hazardous waste, excessive energy use, or air and water pollution during production, use or disposal. Write letters to commend companies that made the best use of their packaging and to criticize those that are wasteful or unsafe.

Ralph Nader, on the 20th anniversary of Earth Day, said Americans have been spending so much time worrying about getting ahead in our careers and in our social lives, that we have forgotten about our civic lives. Meanwhile, people in China and Eastern Europe are putting their lives on the line for the right to have a civic life.

Exercising your civic rights means taking the responsibility to study environmental issues and learn how to jump through the government's hoops to solve the problem. Governments do respond to public pressure. State Representative Mary Brown, D-Kalamazoo, said three well-crafted letters from constituents is enough for her to act on an issue. She is working to strengthen Michigan's air pollution law.

States in the Great Lakes basin have some very forward-looking environmental legislation. But unfortunately, what the government puts down on paper does not always translate into action. Citizens are the key to making these paper promises a reality by voicing their concerns and urging elected officials to implement new environmental programs, strengthen existing laws and make sure that regulations are enforced. Our government depends on people as part of its system of checks and balances. If the public neglects its civic responsibility of keeping the government accountable to the people it was created to serve, our environmental state will stagnate.

Citizens must demand access to better information from industry and the government. What types and amounts of toxic by-products are community corporations producing and how do they dispose of them? How will these substances affect the health of the people who live near the facility?

We need to stimulate better communication between industry and the people who live nearby. Residents can encourage their industrial neighbors to sign a "Good Neighbor Agreement" to surpass minimum pollution requirements.

But civic involvement does not only mean that citizens have to take on big industries and the government. Donate your time and money to an environmental organization. If you are a teacher, work more environmental education into your curriculum. If an environmental problem surfaces in your neighborhood, call attention to it by contacting the local newspaper, radio or TV news station.

Members of service organizations such as the Rotary Club or Junior League can initiate environmental projects. If you own undeveloped land, consider bequeathing it to a group like the Nature Conservancy. Whatever you choose to do, put faith in the power of people working together to improve the quality of their common environment.

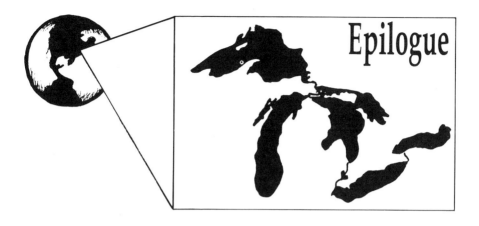

Epilogue

The good news is that more people are worried about the future of the planet than ever before. At the same time that this concern is rising, the cold war is ending. This presents the opportunity for redefining security and for seeing that the future threats are much less military and much more environmental."

This quote from Lester Brown, president of the Worldwatch Institute, aptly predicts that natural dangers such as toxic contamination, acid rain and climate change will be our greatest foes as we head into the 21st Century. It is hoped the nations of the world can act quickly enough to reverse the many environmental problems that threaten the globe before they reach the crisis stage.

Albert Einstein once said, "Smart people solve problems, but geniuses prevent them." We need to find ways to persuade industries to prevent pollution by reducing it at the source instead of searching for increasingly complicated technologies to "dispose" of it.

Much of the pollution control technology used in our country shifts contamination from one part of the environment to another. Many of the state and federal agencies created to deal with pollution also operate in this fragmented manner instead of viewing the ecosystem as a single, interconnected entity.

Compared with other states, Michigan has some exceptional, trend-setting environmental legislation. The state Department of

Natural Resources has also been a national leader in preventing hazardous pollution from new sources, despite a lack of guidance from federal agencies.

But other environmental legislation, such as the Michigan Air Pollution Control Act, are stained with the indelible mark of the state's richest and most powerful industries: the major utility companies, the automakers and chemical companies. So far, Michigan's corporations have learned to control the political process and government regulation better than its average citizens. It's time for the citizens of Michigan to take a more active role in government decisions in order to restore a more equitable balance between the economic health of the state and its environmental well-being.

Perhaps the key to restoring our degraded environment is to live these words: "I give my pledge as an American to save and to faithfully defend from wastes the natural resources of my country—its air, soil, and minerals; its forests, waters and wildlife."

—Conservation pledge

ABOUT THE AUTHOR

Melissa A. Ramsdell, a native of East Lansing, Mich., has devoted much of her writing endeavors to environmental subjects.

A Scripps Howard Foundation scholar, Ramsdell recently earned a master's of science degree from Columbia University's graduate school of journalism in New York. For her thesis on environmental reporting, Ramsdell studied the practice of placing environmentally-threatening facilities in low-income neighborhoods.

Ramsdell also served as an intern at The Morning News Tribune in Tacoma, Wash., and the Toledo Blade in Ohio.

While working on her bachelor's of art in English from the University of Michigan, Ramsdell was city editor of The Michigan Daily, the university's student newspaper.

Ramsdell enjoys skiing, sailing and camping in Michigan.

ABOUT THE ILLUSTRATOR

Jean S. MacKenzie of Lansing, Michigan, holds a master's degree in fine art from Michigan State University. Now a landscape architect, Ms. MacKenzie also paints and sells her oil paintings and prints through a gallery in Grand Rapids.